Fishes: *The Animal Answer Guide*

Fishes

The Animal Answer Guide

Gene Helfman and Bruce Collette

The Johns Hopkins University Press   Baltimore

© 2011 The Johns Hopkins University Press
All rights reserved. Published 2011
Printed in the United States of America on acid-free paper
9 8 7 6 5 4 3 2 1

The Johns Hopkins University Press
2715 North Charles Street
Baltimore, Maryland 21218-4363
www.press.jhu.edu

Library of Congress Cataloging-in-Publication Data

Helfman, Gene S.
    Fishes : the animal answer guide / Gene Helfman and Bruce Collette.
        p.   cm.
    Includes bibliographical references and index.
    ISBN-13: 978-1-4214-0222-2 (hardcover : alk. paper)
    ISBN-10: 1-4214-0222-X (hardcover : alk. paper)
    ISBN-13: 978-1-4214-0223-9 (pbk. : alk. paper)
    ISBN-10: 1-4214-0223-8 (pbk. : alk. paper)
    1. Fishes—Miscellanea. I. Collette, Bruce B. II. Title.
    QL617.H35 2011
    597—dc22   2011008289

A catalog record for this book is available from the British Library.

All photos are by Gene Helfman unless otherwise indicated.

*Special discounts are available for bulk purchases of this book. For more information,
please contact Special Sales at 410-516-6936 or specialsales@press.jhu.edu.*

The Johns Hopkins University Press uses environmentally friendly book
materials, including recycled text paper that is composed of at least 30 percent
post-consumer waste, whenever possible.

All is ichthyological in design, if not intent.

—John Fahey

Contents

Acknowledgments xi
Introduction xiii

1 Introducing Fishes 1
What are fishes? 1
What is the plural of fish? 1
How many kinds of fishes are there? 2
Why are fishes important? 2
What is the most important fish in America? 4
Why should people care about fishes? 4
Where do fishes live? 5
What is the current classification of fishes? 5
Why do we need a system of classification? 6
What is a species? 6
How are species arranged in a classification? 6
What characterizes the major groups of bony fishes? 7
When did fishes first evolve? 8
What is the oldest fossil fish? 9

2 Form and Function of Fishes 10
What are the largest and smallest living fishes? 10
What is the shortest-lived fish? 11
What is the longest-lived fish? 11
Do all fishes have bones? 12
Do all fishes have fins? 13
Do all fishes have teeth? 14
Do all fishes have scales? 15
What is the metabolism of a fish? 17
How do fishes breathe under water? 17
How long can a fish live out of water? 17
Can fishes breathe air? 18
What is a gas bladder? 18
What are lungfishes? 18
Why do some fishes live in salt water and others in
    fresh water? 19
Do fishes sleep? 20
Can fishes see color? 21

Can any fishes fly? 21
What are electric fishes? 22
Can any fishes produce light? 24

3   Fish Colors 26
Why are so many fishes silver? 26
What causes the different colors of fishes? 27
Is there a reason for the color patterns of fishes? 29
What color are a fish's eyes? 32
Do fishes change colors as they grow? 33
Do a fish's colors change in different seasons? 34
Is there much geographic variation in the color of a fish
    species? 35

4   Fish Behavior 37
Are fishes social? 37
Why do fishes form schools? 42
Do fishes fight? 46
Do fishes bite people? 47
How smart are fishes? 48
Do fishes play? 51
Do fishes talk? 52
How do fishes avoid predators? 53

5   Fish Ecology 61
Do fishes migrate? 61
How many fish species live in rivers versus lakes? 64
How many fish species live in the ocean? 65
How far down in the ocean do fishes live? 65
Which geographic regions have the most species of fishes? 65
Are there fishes in the desert? 66
Do fishes live in caves? 67
How do fishes survive the winter? 68
Do fishes get sick? 69
How can you tell if a fish is sick? 70
Are fishes good for the environment? 71

6   Reproduction and Development of Fishes 74
How do fishes reproduce? 74
Do all fishes lay eggs? 75
Why do some fishes lay so many eggs but other fishes lay only
    a few? 76

How long do female fishes hold eggs in their body? 78
Where do fishes lay their eggs? 78
Do fishes lay their eggs at the same time and in the same place
    every year? 79
Do fishes breed only one time per year or once in their
    lives? 80
What is a baby fish called? 80
Are all the eggs in the nest full siblings? 80
How is the sex of a fish determined? 82
Do fishes care for their young? 83
How fast do fishes grow? 84
How can you tell the age of a fish? 85

7   Fish Foods and Feeding 87
What do fishes eat? 87
Do fishes chew their food? 94
How do fishes find food? 96
Are any fishes scavengers? 100
How do fishes eat hard-shelled animals? 101
Do fishes store their food? 101
Do fishes use tools to obtain food? 101

8   Fishes and Humans 103
Do fishes make good pets? 103
What is the best way to take care of a pet fish? 105
Do fishes feel pain? 106
What should I do if I find an injured fish or a fish that looks
    diseased? 107
How can I see fishes in the wild? 107
Should people feed fishes? 108

9   Fish Problems (from a human viewpoint) 110
Are some fishes pests? 110
Can there be too many fishes in a lake or river? 111
Do fishes kill ducks in ponds and other bodies of water? 112
Are fishes dangerous to people or pets? 112
Do fishes have diseases and are they contagious? 114
Is it safe to eat fish? 114
What should I do if I get injured by a fish? 116

Contents

10 Human Problems (from a fish's viewpoint) 117
    Are any fishes endangered? 117
    Will fishes be affected by global warming? 118
    Are fishes affected by pollution? 120
    Why do people hunt and eat fish? 121
    Is there such a thing as fish leather? 122
    Why do so many fishes die at once? 123
    Are boats dangerous for fishes? 123
    How are fishes affected by litter? 124
    What can an ordinary citizen do to help fishes? 125

11 Fishes in Stories and Literature 128
    What roles do fishes play in religion and mythology? 128
    What roles do fishes play in Western religions? 131
    Did early philosophers and naturalists mention fishes in their
       writings? 133
    Are fishes in fairy tales? 134
    What is *gyotaku*? 135
    What roles do fishes play in various cultures? 135
    What roles do fishes play in popular culture? 136
    What roles have fishes played in poetry and other
       literature? 139
    Do fishes have culture? 143

12 "Fishology" 144
    Who studies fishes? 144
    Which species of fishes are best known? 144
    Which species of fishes are least well known? 146
    How do scientists tell fishes apart? 146
    How do you become an ichthyologist? 148

    Appendix A: The Classification of Fishes 149
    Appendix B: Some Organizations That Promote Ichthyology
       and the Conservation of Fishes 161
    Bibliography 163
    Index 171

Acknowledgments

Our answers to many of the questions in this book are based on knowledge gained while writing a college-level ichthyology textbook, *The Diversity of Fishes*. We are therefore grateful to our coauthors in that project, Doug Facey of Saint Michael's College and Brian Bowen of the University of Hawaii for their efforts in that venture. Our work here would have been much more laborious without their help. In many ways, they are also coauthors of this guide.

We remain indebted to our professional colleagues and mentors to whom we have turned innumerable times for answers both ichthyological and more general: George Barlow, J. B. Heiser, Dave Johnson, William McFarland, Lynne Parenti, Jack Randall, Vic Springer, Rich Vari, and Stan Weitzman.

For permission to use illustrations and for providing information on various subjects we thank Cecil Berry, Lauren and Colin Chapman, Joe DeVivo, Bud Freeman, David Hall, Jean-Francois Healias, Malia and Devon Helfmeyer, Zeb Hogan, Mike Horn, Dave Johnson, Michael Kanzler, Stephanie Madsden, Matthew McDavitt, Alexandra Morton, Ted Pietsch, Phil Pister, Rich Pyle, Sandra Reardon, Tyson Roberts, Dave Sumang, Paul Vecsei, Ali Watters, and John Wourms.

Our thanks to Vincent Burke and Jennifer Malat of the Johns Hopkins University Press for trusting us to complete this and tolerating our moments of unreasonableness, usually in good humor. They deserve more cooperative if not compliant authors.

We happily dedicate this work to our wives, Judy Meyer and Sara Collette, in gratitude for their encouragement and unstinting patience in light of other responsibilities that were neglected. We promise to make up for it, soon.

Introduction

Bruce Collette and I have both been fascinated by animals, especially fishes, since we were young. I had my first aquarium before I was 10, and by high school in Van Nuys, California, my bedroom was a chaos of tanks, pumps, heaters, and bubbling noises. With encouragement from my biology teachers, a couple of other fellow fish nuts and I started a shark research group. We collected spiny dogfish for a professor at UCLA who was interested in blood chemistry, spending all-nighters on fishing barges off Redondo Beach, catching sharks for science. We were on top of the world.

In college at UC Berkeley, I got more serious about fishes, learned to scuba dive, and managed the Zoology Department's fish collection. I took ichthyology from the late Dr. George Barlow, an animal behaviorist, which got me started observing fish behavior. Upon graduating, I joined the Peace Corps in Palau, Western Caroline Islands, as a fisheries specialist. Palau has spectacular coral reefs and I spent most work hours and every weekend diving, catching, and watching fishes.

After graduate school at the University of Hawaii and Cornell University (lots more fish watching for science), I was hired at the University of Georgia in Athens, as an ichthyologist, where I taught Ichthyology, Animal Behavior, and Conservation Biology for 30 years. I study marine and freshwater fishes, mostly by diving and observing predator-prey interactions. In recent years I've focused on conservation because many of the places I visited early in my career have deteriorated, as have their fishes.

Bruce Collette's interest in animals began during summers at a camp in the Adirondack Mountains. In high school on Long Island, he experimented with color change in frogs; during vacations from his undergraduate days at Cornell University, he studied lizard behavior and morphology while visiting his parents in Cuba. Bruce switched to fishes in graduate school at Cornell and received his Ph.D. for a taxonomic study of a large group of freshwater fishes called darters that live in streams and rivers from Maine to Florida.

In 1960, Bruce accepted a position as an ichthyologist at the National Systematics Laboratory in what is now the National Marine Fisheries Service of the National Oceanographic and Atmospheric Agency, housed in the National Museum of Natural History of the Smithsonian Institution in Washington, D.C.

Bruce's research focuses on the anatomy, systematics, evolution, and biogeography of tunas and their relatives plus other fishes such as halfbeaks,

Gene Helfman learning about Colorado River fishes at the Sonoran Desert Museum, Tucson, Arizona.
Photo by J. L. Meyer

Bruce Collette shows an Atlantic Bluefin Tuna skull to a young visitor to the Sant Ocean Hall of the Smithsonian Institution. Photo by Mike Vecchione

needlefishes, and toadfishes. His research entails visiting major fish collections around the world, collecting expeditions on various vessels, and using scuba to collect and observe fishes. Results of his research have been published in over 250 papers in many scientific journals plus two regional fish guides, *The Fishes of Bermuda* and *Bigelow and Schroeder's Fishes of the Gulf of Maine*. Bruce has also taught ichthyology as a summer field course in Massachusetts, Bermuda, and Maine.

Bruce and I have known each other for probably 25 years. Given our very different research specialties but our common passion for fishes, we cover much of the broad topic of ichthyology. In 1997 we published an ichthyology textbook, *The Diversity of Fishes: Biology, Evolution, and Ecology*, now in its second (2009) edition and the most widely used college-level ichthyology text in the world. But its approach is too technical for a general audience, so we were delighted when the Johns Hopkins University Press asked us to write a fish book for their Animal Answer Guide series. Fishes are just too interesting and important to be reserved for college students. We would have loved to find a book such as this one when our fascination in fishes was new. We hope this book answers questions and sparks the kind of excitement about these wonderful animals that we have felt for so long.

Gene Helfman

Fishes: *The Animal Answer Guide*

Chapter 1

Introducing Fishes

What are fishes?

While probably everyone thinks they know what a fish is, it turns out to be very difficult to actually define "fish" because of the vast diversity of different species of fishes. Recognizing this diversity, one can define a fish as an aquatic vertebrate that breathes with gills and has limbs in the shape of fins. Several other groups of aquatic animals, such as shellfish (clams and oysters), whales, cuttlefish (squids) and fossil ichthyosaurs (reptiles) are sometimes mistakenly considered to be "fishes" but clearly do not fit the definition. Unlike familiar groups of vertebrates such as birds and mammals that can trace their roots back to some common ancestor, fishes include a tremendous diversity of groups with different ancestries. Hence our definition encompasses lancelets, hagfishes, lampreys, sharks, lungfishes, and bony fishes, each evolving from a different ancestral group.

What is the plural of fish?

The word "fish" is singular and plural for a single species: one Green Sunfish, two Green Sunfish. Ichthyologists (people who study fishes) use "fishes" to refer to more than one species, four different species of sunfishes, fishes of the Gulf of Maine. When ichthyologists use the common rather than the scientific (Latin) name of a specific fish species, they capitalize the name. Capitalizing common names avoids the problem of understanding a phrase like "green sunfish." Does this mean a sunfish that is green or does it refer to the Green Sunfish, *Lepomis cyanellus*? This "rule" about capitalization only applies to discussing a single species. For example, salmon species

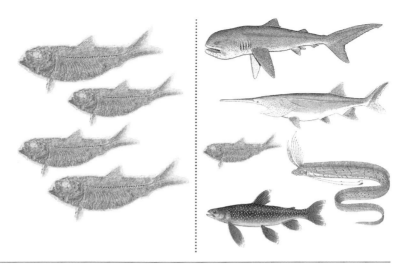

"Fish" versus "fishes": Ichthyologists refer to one species using the singular "fish," regardless of the number of individuals (*left*). When more than one species is discussed, the correct term is "fishes" (*right*). From Helfman et al. 2009; used with permission of Wiley-Blackwell

along the U.S. west coast are referred to as Pacific salmons (lowercase *s*), but each species is capitalized (Chinook Salmon, Sockeye Salmon).

How many kinds of fishes are there?

There are three major groups of fishes: about 115 species of jawless fishes (Agnatha, lampreys and hagfishes), almost 1,300 species of cartilaginous fishes (Chondrichthyes, sharks, skates and rays, and chimaeras), and the remaining 31,000-plus species, the bony fishes (Teleostomi) on which this book will focus (and to which are added more than 200 newly discovered species each year). Bony fishes differ from jawless fishes in possessing true jaws and from both other groups in having a bony instead of a cartilaginous skeleton.

Why are fishes important?

Fishes are most important to people as food. Many families of fishes contain species that form valuable commercial fisheries, such as the Clupeidae (herrings), Engraulidae (anchovies), Salmonidae (salmons), Gadidae (cods), Serranidae (sea basses), Xiphiidae (Swordfish), Scombridae (mackerels and tunas), Pleuronectidae and other families of flatfishes. Other families contain species of great interest to marine sportfishers, such as Coryphaenidae (dolphinfishes or mahi mahis), Istiophoridae (marlins and spearfishes), and Scombridae (tunas), or to freshwater sportfishers, such as Salmonidae (trouts and salmons), Esocidae (pike and pickerels), and Centrarchidae (black basses and sunfishes).

Other fishes are important in the pet trade. These include freshwater fishes such as the Cyprinidae (barbs, goldfish, and other minnows), Characiformes (several families of tetras and relatives, mostly from South Amer-

Freshwater angelfish (*Pterophyllum scalare*) are South American members of the cichlid family. Cichlids are most diverse in Africa, where hundreds of species occur in major lakes and represent some of the most spectacular examples of rapid evolution in the animal world. Cichlids are popular aquarium fishes, behaving and breeding actively in captivity.

ica), Siluriformes (several families of catfishes from South America, Africa, and Asia), Cichlidae (cichlids), and Anabantoidei (Siamese fighting fish and gouramis); and marine fishes such as Pomacentridae (damselfishes), Acanthuridae (tangs, surgeonfishes), Labridae (wrasses), Gobiidae (gobies), and Scorpaenidae (lionfishes).

Because of their small size, ease of care, rapid growth, and short generation time, fishes such as the Zebrafish (*Danio rerio*, Cyprinidae) and the Medaka (*Oryzias latipes*, Adrianichthyidae) are becoming increasingly important in studies of toxicology, pharmacology, neurobiology, and medical research.

Even the gas bladders of some fishes have uses. These organs, also called swim bladders, help fishes maintain buoyancy (see "What is a gas bladder?" in chapter 2). Those from fishes such as sturgeons (Acipenseridae) have been used to prepare isinglass, a whitish, semitransparent, very pure gelatin valued as a clarifying agent and as a component of glue and jellies. Gas bladders from other fishes, such as drums (Sciaenidae), are used in soups.

Perhaps the strangest use of fishes is for pedicures, using small fishes to nibble away flakes of dead skin from peoples' feet. The use of fish to treat skin conditions is popular in Turkey where warm pools are stocked with a small minnow called *Garra rufa*. Live fish pedicures in the United States have used juvenile tilapia (Cichlidae). However, the Texas Department of Licensing and Regulation has banned use of live fish for pedicures because of a worry that the fish might spread infection from one person's foot to another's.

The Zebrafish (*Danio rerio*) is a tropical freshwater fish in the minnow family Cyprinidae. Native to streams in the Himalayan region of Asia, it is a popular and lively aquarium fish and has become the fish equivalent of the white rat for laboratory studies. Photo by Azul

What is the most important fish in America?

Perhaps the most important fish in America are the menhadens (four species of *Brevoortia*, Clupeidae). Menhaden have played an integral if unheralded role in America's history. Native Americans taught the Pilgrims to plant menhaden with their corn. This fish made agriculture viable in the eighteenth and nineteenth centuries for those farming the rocky soils of New England and Long Island. Menhaden fisheries were one of America's largest industries in the nineteenth century. The annual haul of menhaden weighs more than the combined commercial landings of all other finfish. However, you will not see any menhaden on sale in your local fish markets. Where do all the menhaden go? Menhaden are "reduced" into oil, solids, and meal. The oil is used in cosmetics, linoleum, health food supplements, lubricants, margarine, soap, insecticides, and paints. The dried-out carcasses are then pulverized and shipped out as feed for cats and dogs, farmed fish, and, most of all, poultry and pigs. Most of the menhaden are caught and processed by Omega Protein, which has 61 ships, 32 spotter planes, and five production facilities in Virginia, Louisiana, and Mississippi.

Menhaden are also crucial parts of the diet of sport and commercial fishes—bluefin tunas, Atlantic Cod, Haddock, Bluefish, Striped Bass, King Mackerel, and many other species. They are also a major component of the diet of many marine birds and mammals, including porpoises and toothed whales. Each adult menhaden filters four gallons of water each minute while removing phytoplankton. Thus schools of thousands of menhaden reduce turbidity and prevent or limit devastating algal blooms.

Why should people care about fishes?

Fishes account for more than half of the 55,000 species of living vertebrates (the others are amphibians, reptiles, birds, and mammals) and are the most successful vertebrates in aquatic habitats. Fishes are the most primitive group of vertebrates, with a fossil record dating back more than 500 million years. They are our distant ancestors; so study of fishes helps put the human species in perspective. They are important for food, recreation,

Menhaden (*Brevoortia*) may be the most important fish, ecologically and economically, in the United States. Menhaden are members of the herring family and support a huge but diminishing fishery in the Gulf of Mexico and along the Atlantic coast. They are also important food for a variety of fishes and seabirds. These menhaden are swimming with their mouths open, filtering tiny floating plants out of the water.

and as pets. Fishes are important parts of aquatic food webs in both freshwater and marine communities. Healthy fish populations indicate healthy aquatic habitats, safe for fishing and swimming. Fishes share this planet with us and have as much right to live here safely as we do.

Where do fishes live?

Fishes live almost everywhere that there is water: freshwater streams and lakes, hot springs, caves, estuaries, and throughout the ocean from ice-covered polar regions to coral reefs and from the surface to the deepest reaches. About 41% of the species of fishes live in freshwater, with the most species in South America and Southeast Asia. Marine fishes constitute 58% of fishes, and their greatest abundance is in the so-called coral triangle, the area from Taiwan and the Philippine Islands south to Indonesia and the island of New Guinea. The remaining 1% of fishes move between fresh and salt water during their life cycles. The altitude record for fishes is held by some river loaches (Balitoridae) that inhabit Tibetan hot springs at elevations of 5,200 meters (17,000 feet). The record for unheated waters is Lake Titicaca in northern South America, where pupfishes (Cyprinodontidae) live at an altitude of 3,812 meters (12,500 feet). The deepest dwelling fishes are cusk-eels that occur down to 8,000 meters (26,000 feet) in the deep sea.

What is the current classification of fishes?

The classification of fishes varies depending on which ichthyologist (fish specialist) you ask. In spite of some continuing controversy, in this book we will generally follow the classification used by Joseph Nelson in the fourth (2006) edition of his *Fishes of the World*, with a few of the changes recently suggested by E. O. Wiley and David Johnson. That classification is summarized in appendix A.

Why do we need a system of classification?

Things must be named and divided into categories before we can talk about them. This includes cars, athletes, books, plants, and animals. We cannot deal with all members of a class, such as the 31,000 species of fishes, individually. We must put them into some sort of classification. Different types of classification are designed for different functions. For example, one can classify automobiles by function (SUV, van, pickup, etc.) or by manufacturer (Ford, General Motors, Toyota, etc.). Baseball players can be classified by position (catcher, pitcher, first baseman, etc.) or by team (Cubs, Orioles, Giants, etc.). Books may be shelved in a library by subject or by author. Similarly, animals can be classified ecologically as grazers, omnivores, carnivores and so forth or phylogenetically, on the basis of their evolutionary relationships.

There are good reasons for ecologists to classify animals ecologically, but this is a special classification for special purposes. The most general classification is considered to be the most natural classification, defined as the classification that best represents the phylogenetic (or evolutionary) history of an animal and its relatives. A phylogenetic classification is predictive. If one species of fish in a genus builds a nest, it is likely that other species in that genus will also do so.

What is a species?

Species are the fundamental unit of classifications. C. Tate Regan, an early twentieth-century British ichthyologist, defined a species as "a community, or a number of related communities whose distinctive morphological characters are, in the opinion of a competent systematist, sufficiently definite to entitle it, or them to a specific name." This practical, although somewhat circular, definition of a species has been refined with modern evolutionary theory and practice.

How are species arranged in a classification?

Ichthyologists use a large number of units to show evolutionary relationships at different levels. Most of these are only useful to a specialist in a particular group, but some are of more general value. For example, ray-finned fishes fall into the following units: kingdom: Animalia; phylum: Chordata; superclass Gnathostomata (jawed vertebrates); grade: Teleos-

Table 1.1. Scientific classification of three common fish species, showing the different levels that taxonomists use to understand relationships

| | Common name | | |
Taxonomic unit	Atlantic Herring	Yellow Perch	Atlantic Mackerel
Division	Teleostei	Teleostei	Teleostei
Subdivision	Clupeomorpha	Euteleostei	Euteleostei
Order	Clupeiformes	Perciformes	Perciformes
Suborder	Clupeoidei	Percoidei	Scombroidei
Family	Clupeidae	Percidae	Scombridae
Subfamily	Clupeinae	Percinae	Scombrinae
Tribe	Clupeini	Percini	Scombrini
Genus	*Clupea*	*Perca*	*Scomber*
Species	*harengus*	*flavescens*	*scombrus*
Subspecies	*harengus*		
Author	Linnaeus	Mitchill	Linnaeus

Note: Group names are formed from a stem, such as *Clupe-*, *Perc-*, and *Scomb-*, for these three species plus uniform endings for order (-*iformes*), suborder (-*oidei*), family (-*idae*), subfamily (-*inae*), and tribe (-*ini*). The scientific name of a species is made up of the genus and species terms. The author is the person who originally assigned a species its scientific name.

tomi or Osteichthyes (bony fishes); and class: Actinopterygii (ray-finned fishes). Classification of three common fishes down to the next levels is shown in the table in this chapter.

Note the uniform endings for order (-iformes), suborder (-oidei), family (-idae), subfamily (-inae) and tribe (-ini). A group name is formed from a stem, like perc-, plus one of these endings. Thus the Yellow Perch, *Perca flavescens*, belongs to the tribe Percini, subfamily Percinae, family Percidae, suborder Percoidei, and order Perciformes. Each group, going from tribe to order, contains more and more related species.

What characterizes the major groups of bony fishes?

There are two major groups of bony fishes (Teleostomi): the class Sarcopterygii, or lobe-finned fishes, which includes only six species of lungfishes and both species of coelacanths (*Latimeria*), and the class Actinopterygii, or ray-finned fishes, which includes all the rest. Recent studies using several lines of evidence have determined that sarcopterygians also include the subclass of tetrapods, which is the taxonomic designation for amphibians, reptiles, birds, and mammals. Since humans are mammals, that makes us fish. Really.

The African Coelacanth (*Latimeria chalumnae*, Latimeriidae) is considered a living fossil. Coelacanths were thought to have gone extinct with the dinosaurs, 65 million years ago. One fish was caught off South Africa in 1938, and a small population was discovered in 1954 off the Comoros Islands near Madagascar. Many fish were then captured for museums and the population declined rapidly. Coelacanths are now protected internationally from all fishing, and more populations have been found off eastern Africa and Indonesia. This is a life-size replica of a newborn coelacanth.

The Actinopterygii includes several families of mostly freshwater primitive fishes such as the African bichirs (Polypteridae), paddlefishes (Polyodontidae), sturgeons (Acipenseridae), gars (Lepisosteidae), and the North American Bowfin (Amiidae), as well as the more advanced teleostean fishes.

Modern bony fishes are usually placed in the Teleostei (which means "perfect bone" in reference to their strutted skeletal architecture). The Teleostei contains four subdivisions: (1) Osteoglossomorpha, which is made up of the primitive freshwater bonytongues; (2) Elopomorpha, made up of tarpons, bonefishes, and eels; (3) Otocephala, with two superorders, Clupeomorpha (herrings and anchovies) and Ostariophysi (the dominant group of freshwater fishes—minnows, suckers, characins, loaches, and catfishes); and (4) Euteleostei, the advanced bony fishes. About nine superorders of Euteleostei are usually recognized, including the Esociformes (pickerels and mudminnows); six superorders of marine, mostly deep-sea fishes; Paracanthopterygii (cods, toadfishes, anglerfishes); and Acanthopterygii, typical spiny-rayed fishes. A list of the orders and families of bony fishes, including the species mentioned in the text, is included in Appendix A.

When did fishes first evolve?

The very first fishlike vertebrates undoubtedly evolved from invertebrates, perhaps from a cephalochordate similar to today's lancelet, or amphioxus. The first fish fossil appeared in the Early Cambrian, roughly 530 million years before present. One major group of early fishes was once-called ostracoderms (or "shell-skinned") in reference to a bony shield that covered the head and thorax. These fishes dominated the Paleozoic Era

Fishes: The Animal Answer Guide

The Bowfin (*Amia calva*) is the sole living member of its family and occurs only in North America. The fish is an inhabitant of swampy regions and large, slow rivers, feeding at night on a variety of other fishes. It moves silently and slowly by passing waves down its long dorsal fin.

in the Ordovician, Silurian, and Devonian periods, 400–500 million years before present. The Devonian period is sometimes referred to as the Age of Fishes because of the great variety of fishes that occurred then, most of which are now extinct.

What is the oldest fossil fish?

Among the currently recognized oldest fish species was *Myllokunmingia fengjiaoa*, found in the Chengjiang geological formation of Yunnan Province in southwestern China in deposits calculated to be about 525 million years old. It was 3–4 centimeters (1.2–1.6 inches) long and is thought to be an ancestor of modern lampreys.

Chapter 2

Form and Function of Fishes

What are the largest and smallest living fishes?

Body length ranges more than a thousand-fold in fishes. The largest living species of fish is a shark, the Whale Shark, *Rhincodon typus* (in its own family, Rhincodontidae). Whale sharks reach at least 12 meters (about 40 feet) in length and 12,000 kilograms (about 26,000 pounds) in weight but are known to grow much larger, perhaps as large as 18 meters (60 feet) and 34,000 kilograms (75,000 pounds). Next largest is another shark, the Basking Shark, *Cetorhinus maximus* (also in its own family, Cetorhinidae) that reaches perhaps 15.2 meters (50 feet) and 4,000 kilograms (8,800 pounds). The longest bony fish is the 8 meter (26 feet) or longer Oarfish, *Regalecus glesne* (Regalecidae). Large marine bony fishes include two billfishes (Istiophoridae) and a tuna (Scombridae) sought by big-game fishermen, the Black Marlin, *Istiompax indica*, for which there is a gamefish record of 708 kilograms (1,560 pounds), the Blue Marlin (*Makaira nigricans*) at 635 kilograms (1,400 pounds), and the Atlantic Bluefin Tuna (*Thunnus thynnus*), 679 kilograms (1,496 pounds). The Ocean Sunfish, *Mola mola* (Molidae) is much shorter than the billfishes and tuna, 3.6 meters (12 feet), but heavier, 2,300 kilograms (5,000 pounds).

The largest freshwater fishes include the Beluga Sturgeon, *Huso huso* (Acipenseridae), found in Eastern Europe and Asia, which reaches 8.6 meters (28 feet) and a weight of 1,300 kilograms (2,900 pounds), and the Arapaima or Pirarucu of South America (*Arapaima gigas*, Osteoglossidae) that reaches 2.5 meters (8 feet).

The smallest fish—and smallest vertebrate—is a goby (Gobiidae). An Indian Ocean goby, *Trimmaton nanus* and several other gobioids in the gen-

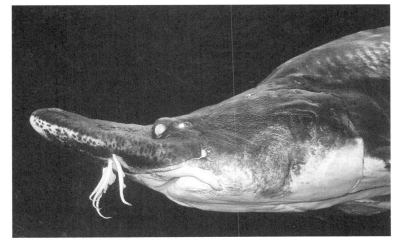

The Russian Beluga Sturgeon, *Huso huso*, is probably the largest freshwater fish in the world. Belugas have been drastically overfished for their valuable eggs. The resulting beluga caviar can sell for over $165 an ounce ($5,280/kg), making a very large beluga worth more than a million dollars.

era *Eviota*, *Mistichthys*, *Pandaka*, and *Schindleria* mature at 8–10 millimeters (less than a half inch). In fresh water, an Indonesian minnow, *Paedocypris progenetica* matures at 7-8 millimeters (less than a half inch).

What is the shortest-lived fish?

The pygmy coral reef goby *Eviota sigillata* is the shortest-living vertebrate known, with a maximum recorded age of 59 days and a maximum length of about 18 millimeters (0.75 inch). Two other species in the same genus are slightly larger, 25–27 millimeters (about an inch) and also have short life spans, less than 100 days. Pygmy gobies grow rapidly, produce several generations over their exceedingly brief lifespan, and experience high daily mortality.

Two groups of fishes, the South American and African rivulid killifishes (Aplocheilidae and Nothobranchidae) are called annual fishes because they live less than a year. These fishes live in temporary ponds that dry up year after year leaving them no water in which to live. As the water level goes down, they lay their eggs in the mud. The eggs survive in the damp mud until the rains come again, hopefully the next year. Then the eggs hatch, and the newborn fish live out their brief life span. This ability to survive as eggs in a damp medium is used by aquarists to ship eggs to other aquarists around the world.

What is the longest-lived fish?

Sturgeons (Acipenseridae) and some species of scorpaenid rockfishes (Scorpaenidae) can live as long as 150 years. At least five species of Pacific coast rockfishes in the genus *Sebastes* can live more than 100 years, the

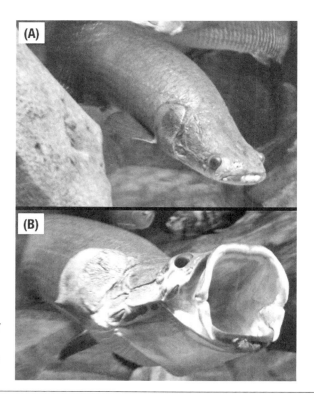

(A), The Arapaima, or Pirarucu (*Arapaima gigas*, Osteoglossidae), is South America's largest freshwater fish. Arapaima have been stocked in lakes and reservoirs in Southeast Asia, where they are actively sought as sport fish. (B), This approximately 1 meter (3 feet) fish demonstrates how its mouth expands to allow it to eat fairly large prey fishes.

Canary Rockfish (*Sebastes pinniger*) for 84 years, and 22 other species for more than 50 years. The record may be held by the Rougheye Rockfish (*Sebastes aleutianus*), which can reach 205 years. Even the Spiny Dogfish Shark (*Squalus acanthias*) can live for 100 years. Long life spans like these have important practical implications, because if fisheries remove too many fish, it may take more than a century for the population to recover.

Do all fishes have bones?

Living representatives of two of the major groups of fishes, the jawless fishes (hagfishes, Myxinidae; lampreys, Petromyzontidae) and cartilaginous fishes (Chondrichthyes), the sharks, rays, and chimaeras, lack true bone, but all the bony fishes (Osteichthyes) have bones. However, most of the skeleton of the lungfishes, coelacanths, sturgeons, and paddlefishes, all members of evolutionarily primitive groups, is made of cartilage. The osteology (study of bones) of fishes is more complicated than in other vertebrates because fish skeletons have many more bones. For example we have only 28 different bones in our heads, a primitive reptile has 72, and a primitive fossil fish more than 150 skull bones. There has been a general evolutionary trend toward fusion and reduction in number of bones in the skeleton.

Fishes: The Animal Answer Guide

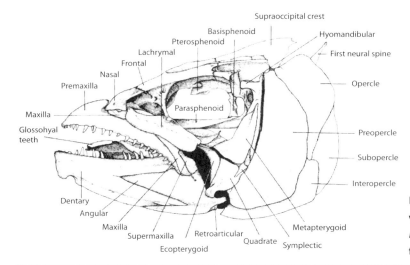

Supraoccipital crest
Basisphenoid
Pterosphenoid
Lachrymal
Frontal
Nasal
Premaxilla
Maxilla
Glossohyal
teeth
Parasphenoid
Hyomandibular
First neural spine
Opercle
Preopercle
Subopercle
Interopercle
Dentary
Angular
Maxilla
Supermaxilla
Retroarticular
Quadrate
Metapterygoid
Ecopterygoid
Symplectic

Bony fishes are *very* bony. This side view of a Dogtooth Tuna (*Gymnosarda unicolor*) skull shows the technical names of many bones.

Do all fishes have fins?

Most fishes have two sets of fins, median or unpaired fins and paired fins. The median fins are the dorsal, anal, caudal, and adipose fins along the dorsal (top), posterior (rear end), and ventral (belly) regions of the fish. The paired fins are the pectoral and pelvic fins, parallel to our arms and legs. Bony fishes have fin rays called "lepidotrichia," bony supporting elements in the fins that are derived evolutionarily from scales.

Primitive soft-rayed fishes, such as herrings, minnows, trouts, and lanternfishes, have a single dorsal fin that is composed entirely of soft rays. Some cods (Gadidae) have three dorsal and two anal fins. Advanced spiny-rayed fishes, such as perches, basses, and tunas, usually have two dorsal fins, an anterior first dorsal fin composed of sharp spines and a posterior second dorsal fin composed of soft rays (although there may be one spine at the anterior end). The second dorsal and anal fins may be followed by a series of small finlets in mackerels and tunas (Scombridae). The anal fin usually lies just posterior to the anus. In soft-rayed fishes it is composed of soft rays like the single dorsal fin of these fishes. In spiny-rayed fishes, the anal fin usually contains one or several anterior spines, followed by soft rays.

Five groups of soft-rayed fishes have an additional fin posterior to the dorsal fin, the so-called adipose fin. This is a poor name for this fin because it is rarely fatty. It usually lacks any supporting fin rays, and its functions are something of a mystery. However, it is useful in distinguishing these five groups of fishes: characins (Characiformes), catfishes (Siluriformes), trouts and salmons (Salmoniformes), lanternfishes (Myctophiformes), and trout-perches (Percopsidae).

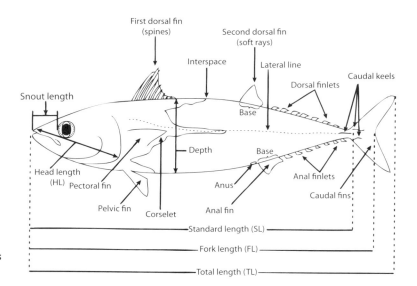

First dorsal fin (spines)

Interspace

Second dorsal fin (soft rays)

Lateral line

Caudal keels

Dorsal finlets

Snout length

Base

Head length (HL)

Depth

Base

Pectoral fin

Anus

Anal finlets

Pelvic fin

Corselet

Anal fin

Caudal fins

Standard length (SL)

Fork length (FL)

Total length (TL)

A mackerel (Scombridae) with the names of its fins and examples of measurements used by ichthyologists in comparing different species of fishes.

The tails of most fishes have a caudal fin that is important in providing power for swimming. However, several groups of elongate fishes, such as many eels, have the caudal fin either greatly reduced or totally absent.

Do all fishes have teeth?

Most fishes have teeth on their upper and lower jaws. The shape and size of the teeth are related to the type of food that a fish eats. Fish and squid eaters have sharp-pointed or triangular teeth useful for grasping slippery prey (e.g., Bowfin, deep-sea anglerfishes, goosefishes), with many teeth curved backward to prevent escape (e.g., Northern Pike, Esocidae). Sharp cutting teeth are common in sharks but much less common in bony fishes, although some South American characins, such as *Myleus*, have cutting teeth for eating plants, as do the infamous piranhas (*Serrasalmus*, *Pygocentrus*) and the barracudas (Sphyraenidae) for cutting flesh. Insect eaters have fine, pointed teeth (sunfishes, Centrarchidae), leaf choppers and algae-scrapers have chisel-like teeth or hardened jaw ridges (cichlids, minnows), and mollusk eaters have molar-like teeth (Sheepshead, Sparidae; wolffishes, Anarhichadidae). Some fishes have their teeth fused into beaks for scraping algae off corals, as in parrotfishes (Labridae), or for biting crustaceans or echinoderms, as in blowfishes (Tetraodontidae).

Unlike most other vertebrate groups, fishes may also have teeth in the roof of their mouth or on their tongue, and some have teeth only on these other mouthparts. Some fishes have another set of teeth called "pharyngeal," or throat, teeth that are located farther back in the head, just in front of the

Fishes: The Animal Answer Guide

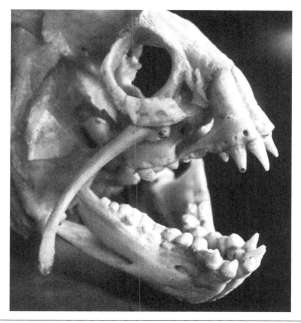

Wolffishes, such as this Wolf-eel, *Anarrhichthys ocellatus*, have teeth more like those of mammals than like the teeth of other fishes. They possess biting and grasping teeth in the front of their jaws and crushing teeth toward the back, analogous to our canines and molars respectively.

esophagus. These upper and lower pharyngeal jaws are well developed in some fishes that also have well-developed teeth in their mouths, such as cichlids and parrotfishes. But they are the only teeth found in the large freshwater family of minnows and carps (Cyprinidae) and in the suckers (Catostomidae).

Do all fishes have scales?

Most fishes are covered with scales that form a protective boundary between the fish's body and the external environment. Scales are of four basic types. "Placoid" scales are characteristic of cartilaginous fishes and are very similar in structure to our teeth. "Cosmoid" scales were present in fossil coelacanths and fossil lungfishes and probably arose from fusion of placoid scales. "Ganoid" scales were present in primitive ray-finned fishes and are found in some living primitive fishes such as sturgeons (Acipenseridae) and paddlefishes (Polyodontidae). Most living bony fishes have either "cycloid" or "ctenoid" scales. Both types of scales evolved from ganoid scales by thinning. These scales overlap like shingles on a roof, which gives great flexibility and less weight compared with the heavy cosmoid and ganoid scales of more primitive fishes. Ctenoid scales have little teeth, called "ctenii," on the posterior margin of the scale. These teeth are absent in cycloid scales. Ctenoid scales are characteristic of more advanced fishes than are cycloid scales.

Scale size varies greatly among fishes. Scales may be microscopic and embedded in the skin as in freshwater eels (Anguillidae), which led to their being

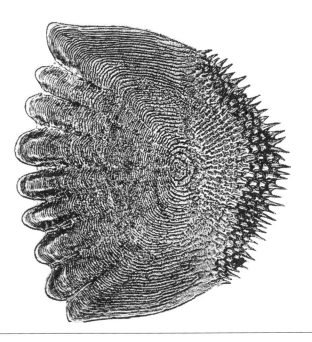

A ctenoid (pronounced "teenoid") scale of a Yellow Perch (*Perca flavescens*) showing the little teeth on the right-hand side called "ctenii." The head end of the fish would be to the left. Fish scales overlap one another so only the ctenii are exposed. Ctenoid scales are characteristic of more advanced (more recently evolved) fishes. Scales without ctenii, called cycloid, are found on less advanced fishes.

classified as non-kosher because of the supposed absence of scales, because tradition has determined that only fishes with fins and scales are appropriate for observing a kosher diet. Similarly, arguments have raged for many years as to whether Swordfish (*Xiphias gladius*) are kosher. While juvenile Swordfish have well-developed but highly modified scales, the scales become deeply embedded in adults so that they appear to lack scales. Scales are small in mackerels (*Scomber*), "normal" in perches (*Perca*), large enough to be used for junk jewelry in Tarpon (*Megalops*), or huge, the size of the palm of a human hand, in the Indian Mahseer (*Tor putitora*), a freshwater cyprinid gamefish reaching 43 kilograms (95 pounds) in weight.

Many fishes show various modifications of scales. There may be an external dermal skeleton in addition to the internal supporting skeleton. This external armor is composed of segmented bony plates in pipefishes and seahorses (Syngnathidae) and poachers (Agonidae). Bony shields similar to placoid scales are found in several South American freshwater families of catfishes that are popular aquarium fishes, such as the Loricariidae and the many species of *Corydoras* (Callichthyidae). The body of shrimpfishes (Centriscidae) is enclosed in bony armor. Trunkfishes (Ostraciidae) are completely enclosed in a rigid bony box with only the fins protruding.

Even though scales are effective in protecting fishes, several groups of fishes have lost their scales over the course of evolution. North American freshwater catfishes (Ictaluridae) have relatively thick, leathery skin. The Ocean Sunfish (*Mola mola*) has the skin reinforced by a hard cartilage layer.

Fishes: The Animal Answer Guide

Snailfishes (*Liparis*, Liparidae) have a transparent jellylike substance up to 25 millimeters (1 inch) thick in their skin.

What is the metabolism of a fish?

Metabolism is the sum total of all the biochemical processes taking place within an organism. Metabolic rates are influenced by a variety of factors, including age, sex, reproductive status, food, physiological stress, activity, season, and temperature. In fishes, the rate of oxygen consumption is frequently used as an indicator of metabolic rate.

Fishes, like other animals, require oxygen to produce enough energy to support their metabolic needs. However, obtaining oxygen from the water is more difficult than it is for us and other air-breathers. Water contains only about 1% oxygen compared with over 20% in air. Gas solubility in water decreases with increasing temperature, so warm water contains less oxygen than cold water. Fresh water can contain more oxygen than sea water (see "How do fishes breathe under water?" below).

Fishes, amphibians, and reptiles are generally considered to be "poikilothermous" (cold-blooded), compared with "homeothermous" (warm-blooded) birds and mammals. Better terms are "ectothermic" and "endothermic," meaning that body temperatures of the first three groups of animals are generally determined by the external temperature of the environment, whereas birds and mammals have a higher metabolic rate and generally maintain stable body temperatures.

How do fishes breathe under water?

Fishes use their gills to obtain oxygen from water. Fish gills are very efficient at extracting oxygen from water, because of their large surface area and the thin membranes of the gill filaments (the bright red structures that you see when you lift up the gill covers of a fresh fish). Also the blood flow in the gill filaments goes in the opposite direction from the water flowing through the fish's mouth. This countercurrent flow ensures that the blood flowing through the gill filaments picks up the maximum amount of oxygen.

How long can a fish live out of water?

Most fishes will die shortly after they are removed from the water. Their gills collapse in air and can no longer gather oxygen. However, some fishes breathe out of water. The Walking Catfish (*Clarias batrachus*, Clariidae) that has invaded Florida has treelike extensions above the second and fourth gill arches that allow them to extract oxygen from air so they can survive for several days out of water. The so-called climbing perches (Anabantidae) of Africa and Asia have a complex folded breathing structure located above the gills that allows them to continue to breathe out of water. This ability to survive out of water allows these fishes to be sold alive in fish markets and transported home for dinner. Snakeheads (Channidae) can also breathe air, and this has allowed them to spread from introductions in Florida and Maryland.

Can fishes breathe air?

At least 370 species of fishes in 49 different families have some ability to obtain oxygen from the air. Most of these fishes only supplement gill respiration when necessary. Lungfishes (subclass Dipnoi) and some other species, such as popular aquarium fishes of the suborder Anabantoidei, like the Siamese Fighting Fish (*Betta splendens*, Anabantidae) and the various species of gouramis, have a specialized organ in their head that allows them to breathe air. These fishes must have access to air or they will drown. This ability to breathe air, and the fact that male Siamese Fighting Fish will attack each other, is why you see male Fighting Fish in separate small jars in aquarium stores.

What is a gas bladder?

The gas bladder, or swim bladder, is a gas-filled sac located between the digestive tract and the kidneys. It is filled with carbon dioxide, oxygen, and nitrogen in different proportions than occur in air, making the term "air bladder" inappropriate. The original function of the gas bladder was probably as a lung in primitive lungfishes. However, in most living fishes, the gas bladder functions mainly as a hydrostatic organ that helps control buoyancy, allowing fishes to hover in mid-water. It also plays a role in respiration, sound production, and sound reception in some fishes. Species in at least 79 of the 425 families of living bony fishes have lost their gas bladders. Many of these are bottom-dwelling species or deep-sea species.

What are lungfishes?

Lungfishes, subclass Dipnoi, are a primitive group of sarcopterygian (lobe-finned) fishes that actually use their gas bladders as a lung in breathing. Although most of the 60 described fossil genera of lungfishes were marine, all the living species occupy freshwater habitats. There are three families of living lungfishes: Ceratodontidae, one species in two rivers of eastern Australia; Lepidosirenidae, one species in South American rivers; and Protopteridae, four African freshwater species. African lungfishes can

Gas bladder of a porcupine pufferfish (Diodontidae). Diodontids have three-chambered gas bladders that function primarily to maintain buoyancy. They inflate themselves when disturbed by pumping water into their greatly enlarged stomachs.

estivate, which means sleep in summer, sort of the opposite of hibernate. They use mucus to coagulate mud to form a mud cocoon around themselves that enables them to survive for up to several years while waiting for the rains to return and to provide them with a proper aquatic habitat. Structure of the heart and lungs confused early (1830s) naturalists as to the relationships of these strange creatures. The South American Lungfish was thought to be some sort of reptile and named *Lepidosiren paradoxa*. An African lungfish, *Protopterus annectens,* was the second species of lungfish to be described and was thought to be an amphibian.

Why do some fishes live in salt water and others in fresh water?

Most families of fishes live primarily in either salt or fresh water. It is difficult to move from fresh into salt water or vice versa because of problems in "osmoregulation," maintaining proper levels of water and salts. Freshwater fishes like minnows (Cyprinidae) and characins (Characiformes) tend to gain water and lose soluble substances across the thin membranes of their gills and pharynx. Marine fishes have the opposite problem. They face dehydration as the high salt concentration of the ocean draws water out of their bodies. Fishes use their gills and kidneys to maintain proper water and salt balance. A relatively few kinds of fishes are "diadromous," meaning they can move from fresh water into salt or vice versa. These fishes have modified cells in their gills that help them adjust to the changes in salt concentrations in the different environments. There are two kinds of diadromous fishes: anadromous (ocean to river) fishes like salmons (Salmonidae) and river herrings (*Alosa*, Clupeidae) move from the ocean and run up rivers to spawn, and catadromous (river to ocean) fishes like freshwater eels (*Anguilla*, Anguillidae) run downstream and move out into the ocean to spawn.

Do fishes sleep?

Many people think that fishes do not sleep because, since they lack eyelids, they cannot close their eyes. But look at the fishes in a darkened aquarium at night. Many will be resting near the bottom, frequently with a blotchier or much lighter color pattern than the species has during active swimming in the daytime. Sleep occurs when a fish assumes a resting posture for a prolonged period, on the bottom or in some form of shelter, and is relatively insensitive to disturbance. By this definition, many species of fishes do sleep. Some parrotfishes and wrasses (Labridae) secrete a mucous cocoon around themselves at night, perhaps to thwart the highly developed senses of roving nocturnal predators such as moray eels or blood-sucking parasitic invertebrates.

On a coral reef, there is a changeover from daytime to nighttime slightly before sunset. Beginning about an hour before sunset, plankton-feeding fishes (such as some butterflyfishes and damselfishes) descend from the water column, while large herbivores (such as parrotfishes and surgeonfishes) migrate from daytime feeding areas to nighttime resting places. Small fishes enter holes and cracks in the reef, larger fishes shelter under overhangs or in depressions. Beginning about 10 to 15 minutes after sunset, the level of activity around the reef drops off precipitously. For nearly half an hour, there is a quiet period when the daytime fishes have gone into hiding and the nighttime fishes have not yet appeared. Then the nighttime fishes seep out of their daytime retreats and move up into the water column and along the reef. Bigeyes (Priacanthidae) and cardinalfishes (Apogonidae) move up into the water column while grunts, snappers, and squirrelfishes (Holocentridae) migrate along predictable paths to nighttime feeding areas over the open sands around the reef. Daytime fishes shelter on and in the reef during the night; nighttime fishes shelter on the reef during the day, often in caves or below overhangs.

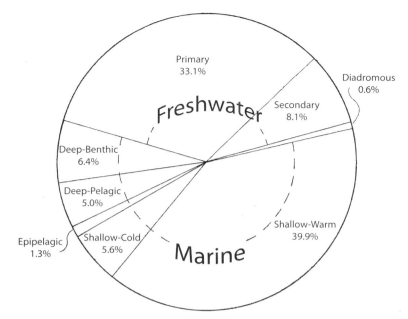

Primary
33.1%

Diadromous
0.6%

Secondary
8.1%

Freshwater

Deep-Benthic
6.4%

Deep-Pelagic
5.0%

Epipelagic
1.3%

Shallow-Cold
5.6%

Marine

Shallow-Warm
39.9%

Percentages of fish species living in different habitats. Primary freshwater fishes are restricted to fresh water; secondary freshwater fishes may venture into marine habitats; diadromous fishes move from fresh water into marine or marine into fresh water at some time during their life cycles. Benthic marine fishes live near the bottom of the ocean, pelagic fishes inhabit the open ocean, and epipelagic fishes are restricted to surface waters.

Adapted from Cohen 1970

Can fishes see color?

Fish eyes are very similar to our eyes except that they have no eyelids. Light passes through a thin transparent cornea and the pupil and then is focused by the lens on the sensitive retina. The retina contains the same two kinds of cells that we have in our eyes. Rods are sensitive to low light levels and are abundant in twilight and nocturnally active fishes as well as deep-sea fishes. Cones are better at detecting motion and provide greater resolution (ability to see very small objects). Cone-to-rod ratios are highest in daytime fishes that rely more on vision. There are several different types of cones, each with a different photoreceptive glycoprotein that responds to different light wavelengths. Many brightly colored shallow-water reef fishes can distinguish different colors, important for finding mates and food and avoiding predators. Freshwater minnows and darters in North American streams may appear pale and silvery or dull brown during most of the year, but in early spring the males change to brilliant colors to attract females for breeding.

Can any fishes fly?

Flyingfishes (Exocoetidae) glide; they do not actually fly. They swim swiftly along just below the surface of the ocean and then launch themselves into the air. The enlarged lower lobe of the caudal (tail) fin is the last part of the fish to leave the water. Flyingfishes can double their speed after emerging into the air, accelerating from about 36 kph (22 mph) in water to as much as 72 kph (45 mph) while airborne. They typically take off into the wind and travel for about 30 seconds and as far as 400 meters in a series of up to 12 flights to escape predators like tunas and dolphin-fishes. They remain aloft for so long by rapidly moving their tail as they descend from a glide. Just the lower lobe of their tail contacts the water, but that's enough to propel them back up to gliding speed. There are two kinds of flyingfishes: two-winged flyingfishes have the pectoral fins enlarged for gliding and four-winged flyingfishes have both the pectoral and pelvic fins enlarged for gliding. While you may think that the fins are moving while a flyingfish is gliding away from your boat, any movement of the fins is just due to air currents. Flyingfishes have no muscles to move their fins up and down so they cannot actually fly.

Flying gurnards (Dactylopteridae) are heavy-bodied fishes with huge pectoral fins that they expand as they "walk" along the bottom. These large pectoral fins have given rise to the idea that they can fly but these fishes are too heavy to move very far off the bottom, much less fly.

However, South American freshwater hatchetfishes (Gasteropelecidae) can fly, sort of. Hatchetfishes have a greatly expanded thorax like a bird does, with muscles attached to their pectoral fins. High-speed photography shows that they move these fins to help them leap out of the water, although the fins are not "flapped" once airborne. Hatchetfishes are popular aquarium fishes but care must be taken to keep a tight lid on the tank because your hatchetfishes are likely to jump out and die on your floor.

What are electric fishes?

All animals produce minute amounts of electricity when their muscles contract or their nerves fire. However, only fishes have developed the ability to utilize electricity for food capture, defense, and navigation. Use of electricity has evolved independently in six groups of fishes: two families of cartilaginous fishes—electric torpedoes (Torpedinidae) and skates (Rajidae)—and four groups of bony fishes—African freshwater elephant-fishes (order Osteoglossiformes), South American freshwater knifefishes (order Gymnotiformes), African freshwater electric catfishes (Malapteruridae) and upside-down catfishes (Mochokidae), and one genus of marine spiny-rayed fishes called electric stargazers (*Astroscopus*, Uranoscopidae).

Electric organs in fishes consist of a large number of disclike plates called "electroplates." One surface receives nervous innervation, and all the electroplates face in the same direction so the current is additive. Each plate is imbedded in a jelly-like matrix enclosed in a connective tissue com-

This two-winged flyingfish (*Exocoetus*) is picking up speed for its glide by sculling with the lower lobe of its tail fin on the water's surface. Flyingfishes can double their escape speed by launching themselves into the air, accelerating to up to 72 kph (45 mph) and thus outdistancing a pursuing predator. www.moc.noaa.gov/mt/las/photos2.htm

partment to provide insulation to the fish. In most cases, the electroplates are derived from muscles: muscle around the gill arches in *Torpedo*, caudal (tail) muscles in skates (Rajidae), eye muscles in *Astroscopus*, and lateral body musculature in most of the others.

There are two types of electric organs in fishes. One, found in four groups of fishes, produces strong stunning current; pulses are produced in small bursts of several milliseconds for offense, defense, and feeding. Electric torpedoes (Torpedinidae) can produce up to 220 volts from large kidney-shaped organs on each pectoral lobe. They use these to stun prey and to repel predators. The South American Electric "Eel" (Gymnotidae, *Electrophorus electricus*) reaches 2 meters (6 feet) in length and can produce 370–550 volts, enough to knock a human down. The word "eel" is in quotation marks because the electric eel is not really a true eel but rather a type of knifefish. The African electric catfish (*Malapterurus*, Malapteruridae) can produce 350–450 volts from lateral body muscles. The only marine bony fish that produces a strong electric current is the electric stargazer, which has modified an eye muscle into an electric organ that can produce about 50 volts. This is thought to be used to stun prey that gets close enough to the fish to be zapped, but unwary fishers that catch a stargazer often get a nasty jolt, suggesting the fish also uses its shocking ability as a defense.

The second type of electric organ produces only a series of low voltage pulses used in detecting objects, navigation, and communication. This

Form and Function

The Electric Eel (*Electrophorus electricus*) produces a weak electric field to locate objects and a strong electric shock to stun prey or repel predators.

type is found along the tail of skates (Rajidae) and in African freshwater elephantfishes (order Osteoglossiformes) and South American freshwater knifefishes (order Gymnotiformes). The electric organs in these fishes sets up a weak electric field around the fish that informs it when the field is disturbed as the fish approaches an object or when another fish swims near it. These three groups of fishes have small eyes, and many of them live in muddy waters where vision is difficult. All these fishes hold their body axis straight while swimming so that the fish does not disturb its own electric field. Elephantfishes have the largest brain of any group of fishes and special receptors in pits on their heads, two characteristics thought to be involved in interpreting electrical impulses. The Electric Eel has both the stunning type and also, typical of the order Gymnotiformes, the location-navigation type.

Can any fishes produce light?

Light production (bioluminescence) is found in fireflies in your backyard but is also common in oceanic invertebrates such as squids, siphonophores, and copepods. Fishes are unique among vertebrates in their ability to produce light. About two-thirds of the approximately one thousand species of deep-sea fishes living in perpetual darkness have light-producing structures called "photophores." Light production has evolved independently in six different lineages: deep-sea dogfish sharks (Squalidae); dragonfishes (eight families in the order Stomiiformes); lanternfishes (about 250 species of Myctophidae); and shallow-water flashlight fishes (Anomalopidae), ponyfishes (*Leiognathus*, Leiognathidae), and toadfishes (*Porichthys*, Batrachoididae). Light can be either autogenic (produced by the fish itself) or symbiotic (produced by bacteria living on or in the fish).

Light is used to tell species and genders apart. Species of dragonfishes (order Stomiiformes) and lanternfishes (Myctophidae) have different, species-specific patterns of photophores along their sides, so the male can find a mate of the proper species in the perpetual dark of the deep sea. Some lanternfishes of the genus *Myctophum* are sexually dimorphic: males have a large luminescent patch on the upper side of their caudal peduncle (area just in front of the tail), whereas females have a smaller patch on the underside of their caudal peduncles. This is how males can tell females apart from males in the dark. Male dragonfishes (Stomiiformes) can distinguish females because males have larger photophores than females in front of, below, or behind their eyes. Lanternfishes such as the headlight fishes (*Diaphus*, Myctophidae) have their snouts covered with luminescent tissue that lights the way in front of them. Flashlight fishes (Anomalopidae) have a large light organ under each eye that they can turn on or off to illuminate their prey of small invertebrates. Flashlight fishes also blink their lights to keep school members together and turn their light on when defending their territory against other flashlight fishes during the breeding season. Dragonfishes have luminous lures at the tip of their long chin barbels, and deep-sea anglerfishes (Ceratioidei) have photophores on their escas, the fleshy bait at the end of their modified first dorsal spines that is used to attract prey.

Chapter 3

Fish Colors

Why are so many fishes silver?

Surprisingly, silvery fishes are silver in order to be invisible. Silver coloration is a characteristic of many fishes that swim up in the water column, often in open water. These fishes—especially the so-called baitfishes such as herrings, minnows, silversides, anchovies—are actually mirror-sided. They have highly reflective crystals in their scales and tend to be very narrow side-to-side, not round or flattened top to bottom.

When viewed from slightly above, we see the dark light that bounces off the fish from below and is reflected into our eyes. We compare it against the dark background of upward traveling light. When we look at such a fish from slightly below, what we see is the bright light coming from above that bounces off the fish and into our eyes, which we compare with the bright background light coming down from the surface. Horizontal light is just that and is the same light directly behind the fish as we look at it from the side. (Some fishes that live over bright sand bottoms are also mirror-sided, such as ponyfishes, Leiognathidae, and mojarras, Gerreidae. When viewed from above, they would reflect the light coming up from the sand and be compared against the bright sand background, so again they would not contrast with the background.)

The sum of all this reflected light is that a mirror-sided fish is the same as a mirror hung in the water, reflecting light that exactly matches the light behind it. As a consequence, the fish looks just like the water background: it disappears. It is the same thing you would get if you hung a mirror, or for that matter a clear pane of glass, in the water. You would not see anything.

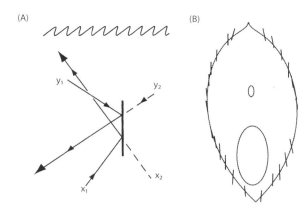

How mirror sides make a fish seem to disappear. (A) Under water, a clear pane of glass is invisible because light coming from behind it passes directly through and into the eyes of an observer (extensions of dashed lines x_2 and y_2 to observer's eyes); an observer sees no difference between the glass plate and its background. A mirror hung in water also disappears because light reflected off the mirror (solid lines x_1 and y_1) is identical to the background light that would pass through the mirror if it were transparent (dashed lines). **(B)** Cross-section through the body of a minnow showing how the reflecting plates under the scales are oriented vertically, even along the curved surfaces of the fish.

Based on Denton and Nicol 1965. Reprinted with the permission of Wiley-Blackwell.

As with any adaptation, mirror sides have costs and drawbacks. For the mirror-sided fish to be invisible, it has to remain upright. Otherwise it would start reflecting brighter or dimmer light than the background against which we compare it. If you sit on a dock and watch a school of baitfish, every now and then you will see a bright flash, as one fish rolls on its side and reflects very bright, downwelling light that contrasts sharply with the dim light below it. A baitfish definitely does not want to call attention to itself, and so these fish tend to swim and hover as upright as possible, with little rolling and turning.

What causes the different colors of fishes?

Fish colors result primarily from pigments (colored substances) in their skin, fins, and scales that reflect and absorb light and from light passing through many layers of skin cells, reflecting the light that has not been filtered out. Yellow, orange, and red are colors reflected by skin pigments. Greens, blues and violets result when light is refracted and reflected by layers of skin and scales; the exact color depends on the number and thickness of the layers. Black usually results from expansion and contraction of special cells containing the pigment melanin, the same compound that causes human skin to darken when exposed to sun (as occurs during tanning).

Silvery sides make a fish invisible, usually. Silvery, or mirror-sided, fishes disappear into the background under natural lighting. But the same reflective characteristics that normally make them invisible also make them conspicuous when light strikes them at abnormal angles. This Cero, a species of Spanish mackerel (*Scomberomorus regalis*) stands out dramatically when illuminated from the side with an electronic flash.

White color is light reflected from cells containing guanine (which is why bird guano is white).

Fishes can change their colors quickly (in seconds) or slowly (as they grow or during different seasons). Short-term color change is usually controlled by the nervous system. Short-term changes occur during behavioral interactions. When two fish fight, they may flash bright colors or intensify dark ones. The winner usually remains darkened while the loser quickly pales. Longer term, growth and seasonal change are more likely controlled by hormone levels. During the breeding season, many fishes, but especially males, take on bright, contrasting patterns. Many fishes go through a series of color changes as they grow. Juveniles in at least 18 families of coral reef fishes have coloration that is different from that of adults.

Even the most colorful fishes by day turn relatively dull or blotchy at night. This day-night difference occurs in just about every habitat. Brilliantly colored butterflyfishes (Chaetodontidae) on coral reefs are blotchy gray at night. Cardinal tetras (*Paracheirodon axelrodi*, Characidae), a popular aquarium fish, are brilliant blue-green and red by day but take on an inconspicuous pinkish tinge as they rest on the bottom at night.

Recent research has revealed an aspect of fish coloration and vision that was previously unsuspected. We have known for a long time that insects can see ultraviolet (UV) light and that many flowers reflect UV light to attract pollinating insects (humans cannot see UV light, but we can get sunburned by it). Many marine and freshwater fishes can in fact see UV, and some fishes reflect UV from their skin in species-specific patterns. Damselfishes (*Dascyllus*, *Pomacentrus*; Pomacentridae) have UV-reflective patches on their fins and faces that are used during communication with other damselfishes. Their predators cannot see UV and so the predators cannot eavesdrop. Some freshwater fishes also use UV reflection for attracting mates.

Fishes: The Animal Answer Guide

Male swordtails (*Xiphophorus*, Poeciliidae) have UV-reflective markings that attract females but that are not visible to the species' main predator, the Mexican tetra (*Astyanax mexicanus*, Characidae).

Many fishes can in fact be identified based on their coloration. It is not only fish biologists that use color to tell fishes apart. The Beau Gregory Damselfish (*Stegastes leucostictus*, Pomacentridae) can apparently tell 50 different species of reef fishes apart based on a combination of body shape and color.

Is there a reason for the color patterns of fishes?

There are two general reasons for the color patterns of fishes. The first has to do with the way light enters and passes through water, creating a visual environment very different from on land. The second, related reason, involves the function of the color. Fishes are colored either to be seen or to not be seen.

Light in water is very different from light on land. All wavelengths (colors) of light are equally available in air across the visible spectrum from short (blue) to long (red) wavelengths of the visible light spectrum. Water transmits light rays differently depending on the wavelength of the light. Long wavelengths in the orange and red are absorbed first, both vertically and horizontally; green and blue travel the farthest. Beyond 10 meter (33 foot) depths, very little red light remains. If you cut yourself at 10 meters, your blood flows green. Similarly, a bleeding object a meter (3 feet) away will appear red but 10 meters away, the blood will appear green.

Also, incoming sunlight is refracted (bent) at the water surface such that 95% of available light is downwelling, very little traveling up or horizontally. In air, horizontal light can be stronger than light coming from the sky above you, as when the sun is low in the sky. All these factors affect the color of fishes.

Fish coloration takes advantage of these properties of underwater light. Most fishes do not want to be visible to their predators or prey. The silvery color of fishes that live in the water column, discussed above, is a good example. Another good example is the many fishes that rest on the bottom or near objects, such as flatfishes, crocodilefishes, scorpionfishes, carpet sharks, toadfishes, darters, gobies, seahorses and seadragons, and sargassumfish. These fishes are generally camouflaged, taking on a mottled color pattern, often with skin growths or fins that look like the twigs, algae, and rocks around them. They mimic and blend in with the background against which they are seen or, more accurately, not seen.

However, most fishes swim actively above the bottom in shallow water. They tend to be countershaded. They are darker on top and gradually lighter on their sides and brightest on their bellies. This gradient of color takes advantage of the progressive decrease in light as measured when looking up, sideways, and down.

Countershading is easiest to understand when viewing a fish from above. A fish with a dark back absorbs bright downwelling light, which creates a dark target against the dark background of dim upwelling light. However, most fishes are viewed by their predators from the side. When viewed from slightly above the horizontal, the dark upper sides of the fish absorb relatively bright downwelling light, creating a dark target that is compared with the dark background of slightly upwelling light. Similarly, if viewed from slightly below the horizontal, the lighter colored side of the fish reflects weak upwelling light, creating a relatively bright target seen against the lighter background of slightly downwelling light.

A countershaded fish thus disappears into the background because the distribution of its color is opposite to the distribution of light in water. This combination of reflected light and fish color creates a target that is identical to the background. The fish reflects light that is roughly equivalent to the background against which it is seen at all viewing angles, dark against dark, intermediate against intermediate, light against light.

But there are times when fishes want to be seen, such as during mating or when fighting over territories. These interactions usually occur when fishes are near one another, and again their coloration takes advantage of the peculiarities of underwater light. Many fishes can turn their colors on and off. Colorful reef fishes often superimpose bright colors over a countershaded body, and then change color patterns depending on whether they are engaged in social interactions or avoiding predators. Hence butterfly-fishes (*Chaetodon*, Chaetodontidae) can be bright or countershaded, and the change can occur in seconds. Male darters (*Percina*, *Etheostoma*; Percidae) in streams are mottled and camouflaged most of the time but take on brilliant blue, orange, and red hues during the breeding season.

Most conspicuous colors, especially the longer wavelengths near the red end of the spectrum, are only visible at short distances or in shallow water. So what appears to be a brilliantly colored fish close up, or when viewed with flash photography, is a dark and dull colored fish when seen from the distance of an approaching predator under natural light. The same color can be advertisement or concealment at the same time, depending on how far away the viewer is from the fish.

The same applies to red or orange deepwater fishes (Orange Roughy, scorpionfishes) hauled to the surface and nocturnal fishes that hide in holes or caves by day (squirrelfishes, Holocentridae; bigeyes, Priacanthidae). When seen under natural lighting in their actual habitat (deepwater fishes) or at the right time (at night when there is little red light available), these fishes are dark and dull where and when it matters.

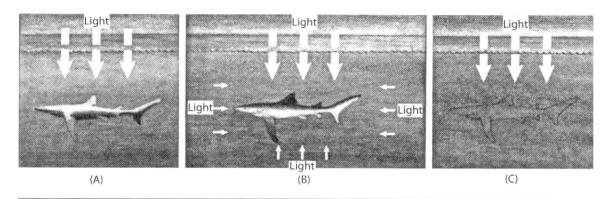

Most fishes are countershaded: darker on top, gradually lighter on their sides, and brightest on their bellies. This color pattern makes them seem to disappear in the water column because their coloration reflects light in a way that matches natural background light. (*A*) A *uniformly gray* fish illuminated from above, as in natural lighting, would have a relatively bright back and a relatively dark belly. (*B*) A *countershaded* fish illuminated from the side, as in flash photography, stands out against the natural light behind it. (*C*) In a *countershaded* fish under natural lighting, the gradual transition from dark back to light belly has an averaging or cancelling effect. The top of the fish is seen as dark against dark, the middle as intermediate brightness against intermediate, and the light belly as light against light. All fish color-background combinations eliminate contrast between the fish and its background. From Helfman et al. 2009; used with permission of Wiley-Blackwell

A related topic deserves mention. This is reverse countershading. Some fishes are dark or colorful on their bellies, grading to lighter on their backs. Reverse countershading would seem to make the fish conspicuous at all angles of viewing, light against dark from above, and dark against light from below. These apparent exceptions however prove the rule that countershading is camouflage. Many male fishes, such as sticklebacks, sunfishes, cichlids, and wrasses, take on bright dorsal or dark ventral colors during the breeding season, a time when conspicuousness helps them attract females and repel territorial intruders. A very good proof-by-exception comes from the reversely countershaded upside-down catfishes (Mochokidae) that feed on the undersides of leaves and even swim in open water while upside down. Predatory Lake Malawi cichlids in the genus *Tyrannochromis* are reversely countershaded and often attack prey while upside down. The name of one species, *T. nigriventer*, means literally "black belly" (one of the upside-down catfishes is similarly called *Synodontis nigriventris*).

The topic of coloration in predators deserves as much attention as the color of their prey. Predators do not want to be detected any more than prey do. Success in predatory fishes is very much dependent on the element of surprise. Prey that recognize a predator and can sense its intentions usually escape. Predators need to be able to sneak up close enough to a prey fish to make a quick, successful dash. Hence anything a predator can do to

Fish Colors

slow the reaction of the prey increases the predator's chances. This is the best explanation for a color pattern common among predatory fish known as split head coloration, a color that only works when you realize that most prey see their predators approaching them head on. Many predators have a contrasting line running between the eyes, from the top of their head down to their mouth. Dark fish have a light line, and light colored fish have a dark line. It is thought that this line splits the profile of the fish's head, a pattern referred to as "disruptive coloration," making the predator harder to recognize, at least for a moment. And a moment is all a predator needs to achieve the crucial element of surprise.

What color are a fish's eyes?

Most fishes have eyes with a white iris surrounding a black pupil. But in some fishes the iris is anything but white. A common pattern is for the color of the head and body to be continued onto the eye. Butterflyfishes among many others often have a black bar on the head that continues through the eye, the iris above and below the pupil also being dusky or black.

Red color in particular is often placed in the iris. Scorpionfishes (*Scorpaena*, Scorpaenidae) have a mottled red iris that matches the mottled red coloration of the head. Some blennies (Blenniidae) have a red longitudinal stripe that runs from the tail base to the snout. This line is continued through the iris and even onto the pupil. In freckle-face blennies (*Istiblennius*), a red-and-green speckling on the body also occurs in the iris. Hawkfishes (Cirrhitidae) often have a red mottled body color, sometimes with green flecks. Their iris is also colored red, sometimes with green flecks. The Red Irish Lord (*Hemilepidotus*, Cottidae) has red mottling on its body, a reddish horizontal band through the iris, and even red speckles in the pupil. Some red-eyed fish however just have red eyes without red bodies. Rock bass and smallmouth bass (*Ambloplites*, *Micropterus*; Centrarchidae) have distinctly red irises but no red on the body or head. The function of red eye color in these fishes is a mystery.

Red and green are not the only colors shared by body and iris. In some butterflyfishes, white or yellow coloration on either side of the head bar is also repeated in the iris. Common Carp (*Cyprinus carpio*, Cyprinidae) and Golden Trout (*Oncorhynchus mykiss aguabonita*, Salmonidae) also have yellow-golden hues in their irises. Australian Blue Eye (*Pseudomugil*, Melanotaeniidae), Blue-Eyed Triplefin (*Notoclinops*, Tripterygiidae) and Blue-Eyed Plecostomus (*Panaque*, Loricariidae) all have blue coloration in their iris. Porcupine pufferfish (*Diodon*, Diodontidae) surround the black iris with a circle of black, and some pufferfishes have black spots on their bodies and similar size black spots in their iris.

This Japanese Akame (*Lates japonicus*, Latidae) has the split head coloration pattern typical of many predatory fishes. A line running between the eyes is thought to disrupt the outline of the predator when viewed by a potential prey victim, making the predator harder to recognize, at least momentarily. A moment usually decides success versus failure during a predator-prey encounter.

Photo courtesy of J. DeVivo

Matching of body and eye color suggests that many fishes are trying to hide their eyes. It is generally felt that eye concealment is necessary because both predator and prey fishes notice eyes. Hence any fish that is trying to avoid detection should conceal as much of its eye as it can. Divers, especially spearfishing divers, learn quickly not to look at fish they intend to spear until the last moment. As soon as you turn your head and face your intended victim, it flees.

Some fishes that live in dimly lit, but not totally dark, water (e.g., stoplight loosejaw, *Malacosteus*, Stomiidae) have yellow lenses. This coloration acts as a filter, cutting out blue-green light and perhaps improving their ability to see the red photophores (light organs) of other loosejaws that emit reddish light.

Do fishes change colors as they grow?

Some fishes remain the same color throughout most of their lives. This is especially true among small, silvery, schooling fishes (silversides, herrings, sandlances, many minnows, sardines). But in many species, juveniles have dramatically different coloration from the adults.

Color change with age is common in fishes that occupy different habitats as they grow. Two good examples are salmon (Salmonidae) and freshwater eels (Anguillidae). Young salmon that have recently hatched from eggs have vertical bars (parr marks) on their bodies, a color pattern that breaks up their body outline and takes advantage of the dappled sunlight typical of the clear, shallow streams in which they first live. When they grow and migrate to the ocean, they change to a uniform silvery color typical of open water, schooling fishes.

American Eels (*Anguilla rostrata*) also change habitats, but they spawn far out at sea. The larvae are carried on ocean currents, at which time the ribbon-like fish are colorless, a good way of being invisible. The larvae then enter coastal streams, rivers, and lakes and become countershaded, which is the baseline color of most fish. When the juveniles mature after several years, they return to the ocean to spawn, at which time they take on a bronze-silvery coloration, again appropriate for an animal swimming in open water.

It is not surprising that many species take on brilliant color during the breeding season and that young and old differ in color as a result of sexual behavior, with males being most colorful. Bluehead Wrasse (*Thalassoma bifasciatum*) are only blue-headed when they achieve the status of territory-holding males. Juveniles, adult females, and adult males that don't hold territories are a less conspicuous yellow overall. But in many coral reef species, the young are much more colorful than the adults. Damselfishes, butterflyfishes, angelfishes, sweetlips, drums, wrasses, and batfishes are among the many families in which this difference occurs. Biologists are at a loss to explain why juveniles—which ought to be more susceptible to predation and thus should be less conspicuous—are strikingly colored and strikingly different from their dull colored adults. Ideas abound: it could be because juveniles are more territorial and use the color to repel intruders, or juveniles could be mimicking toxic invertebrates on the reef, or who knows. This is an interesting and important area for further research.

Do a fish's colors change in different seasons?

Tropical fishes, both in marine and fresh water, tend to maintain the same color pattern throughout the year (some exceptions exist among African cichlids, the males of which intensify their color when breeding). This constant colorfulness is one of the reasons why tropical fishes are popular in the aquarium trade.

In more seasonal regions, such as temperate freshwater lakes and streams of North America, breeding males of many species take on bold and brilliant colors, often reds, blues, yellows, and greens placed next to white or black (minnows, suckers, killifishes, darters, sunfishes, sticklebacks, livebearers, pupfishes, cichlids; see "How do fishes reproduce?" in chapter 6). These fish "color up" as the breeding season approaches, and bigger, more successful males are the brightest. Females in general maintain the same, dull, countershaded or silver color year around.

Fishes: The Animal Answer Guide

Parr color stage of salmon. Young juvenile Atlantic Salmon (*Salmo salar*) have vertical bars on their sides, called "parr marks." This color pattern makes them less obvious in the dappled light characteristic of shallow, clear streams. Adult Atlantic Salmon are silvery, a better color for the light environment of the open ocean. Photo by P. E. Steenstra

Individuals within a population of a species can vary in their color. Great Barracuda, (*Sphyraena barracuda*), often differ in the spot patterns on their sides. The fish may use this variation to help them identify one another.

Is there much geographic variation in the color of a fish species?

Many if not most fish species show little variation in color from place to place. Some cichlids from Africa and Central and South America show geographic variation in color, although sometimes we have discovered that what we thought were different color varieties are in fact different species. North American minnows (e.g., red shiners, yellowfin shiners, *Cyprinella*, *Notropis*; Cyprinidae) often show differences in fin and body color at different locales, usually most obvious in males during the breeding season. African cyprinids (*Barbus*) also show geographic variation in body color. Guppies (*Poecilia reticulata*, Poeciliidae) vary greatly in color from one locale to another for obvious ecological reasons. Fishes from places where daytime predators (wading birds, fishes) occur tend to be less colorful. In places lacking predators, male guppies have the bold and colorful bodies and fins that have made them a favorite of aquarium keepers. Red-bellied piranhas (*Pygocentrus*, Characidae) vary considerably in color at different locales in South America. Many wrasses on Pacific coral reefs also show color variation from locale to locale.

Variation within a species does not just occur on a geographic basis. Individuals on a single reef may also differ. Great Barracuda (*Sphyraena barracuda*, Sphyraenidae) and trumpetfish (*Aulostomus*; Aulostomidae) have patterns of spots on their sides that differ from side to side and from fish to fish. No two fish are exactly alike. For barracuda, at least, the fish themselves may use the spot patterns to help tell individuals apart. Trumpetfish in the Caribbean (*A. maculatus*) carry this individual variation a step further, turning on body coloration that matches the color of their prey. Trumpetfish that feed on blue-colored damselfish take on a blue hue, especially around their heads (the part the prey see when being attacked). Trumpetfish that feed on yellow prey, such as the abundant juvenile bluehead wrasse discussed earlier, have an overall yellow body color. It is thought that individual trumpetfish maintain their matching body color, which is to say blue trumpetfish don't become yellow trumpetfish.

Chapter 4

Fish Behavior

Are fishes social?

Fishes interact socially with members of their own species (conspecifics) and with other species in many ways. Apart from breeding behavior (see chapter 6), fishes interact socially in groups, over territories, around food, and in mutually beneficial relationships with other fishes or animals.

The most common form of sociality that many people associate with fishes is schooling, a topic deserving detailed exploration (see below, "Why do fishes form schools?"). The opposite of group formation and schooling is territoriality, one individual staking out a portion of a reef, kelpbed, or lake bottom and defending it against all comers. Most fishes do not defend territories, some defend only the space around them regardless of where they are, and some actually defend a physically defined location with definite borders. Species in the last group have been well studied because they stay in one place and allow human intruders to get close (but not too close) to make observations.

Territoriality has been observed in a wide variety of fishes, including anguillid (freshwater) eels, minnows, several catfish families, electric knifefishes, many salmons and trouts, frogfishes, sticklebacks, pupfishes, rockfishes, sculpins, sunfishes and black basses, butterflyfishes, cichlids, damselfishes, wrasses, barracudas, blennies, gobies, and surgeonfishes (sharks, hagfishes, and lampreys are apparently not territorial). Territoriality is best known in the damselfishes (Pomacentridae) on coral reefs because they live in warm, clear water (obtaining data underwater requires many hours of observation) and tolerate divers.

In damselfishes, the territory serves multiple purposes, including a food

Male Bluegill Sunfish (*Lepomis macrochirus*) stake out territories and court females. This male Bluegill has dug out a nest over a sandy bottom in a spring in Florida.

resource (algal gardens that the fish actually weeds), a place where eggs are laid and defended in the case of male damselfishes, and holes where the territory holder can hide from predators or rest at night. It is fairly easy to map the territory by simply watching the fish, seeing where it attacks intruders, and dropping weighted markers at the places where these contests occur. The territory is often a well-defined patch of reef about a meter (3 feet) in diameter. Fish passing outside the boundaries are ignored, but any fish crossing over the line is attacked vigorously. Many a diver has been shocked to find an 8 centimeter (3 inch) long damselfish pulling his or her leg hairs or bumping against a face mask until the diver moves back a couple of feet.

Biting is really the last and most desperate tactic employed by the territory holder, and probably occurs because the diver ignored earlier, more subtle displays. If another fish approaches the boundary, the territory holder will swim out and spread its fins and gill covers, swim in place—often in circles—with exaggerated movements, make popping noises with its mouth, change colors, dash at the intruder, and finally, bite. Prolonged contests are usually won by an established territory holder, even if it is smaller than the intruder. If two fish are fighting over a new space, the larger one usually wins.

Territory size differs greatly among different species but generally corresponds to fish size, larger fish having larger territories. Damselfishes, a family that includes the familiar *Amphiprion* anemone or clown fishes, have small territories on the order of a square meter (9 square feet). A clownfish's territory is usually the size of the anemone it occupies with a few other clownfish. At the small extreme are the territories of such fishes as pikeblennies, 12 centimeter (4.5 inch) bottom dwellers that often live in abandoned worm tubes. They defend territories that barely extend outside

Fishes: The Animal Answer Guide

their burrows and are much less than a square meter in size. At the other size extreme are the territories of the Great Barracuda, which can extend along a reef face for almost the length of a football field. In fact, if you go snorkeling in the Florida Keys or many places in the Caribbean, as you swim along a reef you may find yourself being followed by a large barracuda. Basically, you have entered its territory and it is accompanying you until you leave.

As useful as a territory might be in providing food, breeding sites, and refuge holes, owning a territory also involves costs. Territorial defense takes time and energy and carries a risk of injury. A territory holder displaying to an intruder might be less aware of nearby lurking predators. Chasing an intruder from one portion of your territory means other parts are undefended. It is not unusual to see wrasses dash into a damselfish territory and eat eggs while the male damselfish is occupied chasing an algae-munching parrotfish at the territory edge.

As a result of these pluses and minuses, it has been shown that territory holders make what appear to be cost-benefit calculations regarding territory size. A territory holder must obtain all its food within its territory because available nearby food will likely be defended by another fish. A bigger territory might hold more food, but a bigger space is more costly to defend. When researchers add food to a territory, such as a rock covered with algae stolen from another territory, territorial boundaries shrink. This makes sense because now the fish can meet its energy requirements in a smaller area. Or at least that is what has been found when experiments are conducted on rockfishes (Scorpaenidae), surfperches (Embiotocidae), and some damselfishes (Pomacentridae).

But sex, not surprisingly, complicates things. It turns out that male and female Beau Gregory Damselfish (*Stegastes leucostictus*) responded differently to increased food in their territory. Male damselfish did in fact shrink their territory. Females, in contrast, actually increased their territory size when food was added. The conclusion reached by the researchers was that males had some minimum energy requirement that could be met in the smaller territory with its increased food. But female fish make a different calculation. Body size in females is very important to egg production; bigger females can produce more eggs, so anything a female can do to increase her size will be favored (see chapter 6). More food gives a female more energy, which allows her to have a bigger territory, which may mean even more food, continuing the cycle. Thus the benefits of larger territory size outweigh the costs for females but not for males.

Fishes also interact socially with other species. Aside from incidents where one species benefits and the other is harmed—as in predation and parasitism—many such interspecific interactions involve symbiotic (living together) relationships that may be one-sided but with minimal harm to one member of the pair (commensalism), or mutually beneficial, highly evolved pairings (mutualism).

Commensal relationships can be as simple as a fish living upon or inside an invertebrate. Cardinalfishes (Apogonidae) live inside the shells of queen conchs, and pearlfishes (Carapidae) frequently occupy the body cavity of mollusks (clams, oysters) and echinoderms (sea cucumbers, seastars). A famous example, and one that likely explains the origin of the common name for pearlfishes, involves a tropical Pacific species (*Encheliophis*). These pearlfish live between the shells of live blacklip pearl oysters (*Pinctada*) by day and emerge at night to feed. One such pearlfish died inside its host and was entombed there while the oyster laid down layer upon layer of mother-of-pearl, forming what is known as a blister pearl.

Another seemingly commensal relationship involves the well-known interaction between remoras or sharksuckers (Echeneidae) and the sharks, marlins, whales, and turtles to which they attach. Remoras have an evolutionarily modified dorsal fin that creates tremendous suction pressures, allowing the remora to remain attached to its host no matter how fast the host swims. In this manner, the remora gets carried great distances that it alone would never be able to cover and feeds on scraps produced when its superpredator host feeds. The shark suffers a (probably minimal) energy cost associated with having a small hitchhiker attached. Recent studies indicate that sharks and marlins may even benefit from an attending remora. Stomachs of remoras contain copepod parasites from the gills and fins of their hosts, making the relationship more mutualistic than commensal.

Unquestionably mutualistic relationships are known from many fish families, especially among gobies (Gobiidae) that live in close association with sponges, corals, shrimp, and other invertebrates. In goby-shrimp pairs, a blind shrimp continually digs and maintains a burrow in a sandy bottom while the goby sits at the burrow entrance, watching vigilantly for predators. The shrimp, whose activities are reminiscent of a small bulldozer constantly bringing sand out of the burrow and pushing it off to one side, keeps its antennae in physical contact with the goby. If the goby senses danger, it turns and dives into the burrow and the shrimp, sensing the movement, follows immediately.

Two of the best-studied mutualistic relationships involve anemonefish and anemones and cleaning symbiosis by members of the wrasse family (Labridae). Many fishes live among the stinging tentacles of jellyfishes and Portuguese men-of-war (e.g., the Man-of-War Fish, *Nomeus gronovii*, Nomeidae), but the most highly evolved relationship between a fish and a coelenterate involves the pomacentrid anemonefishes (*Amphiprion*, *Premnas*; Pomacentridae) and large sea anemones in the tropical Pacific and In-

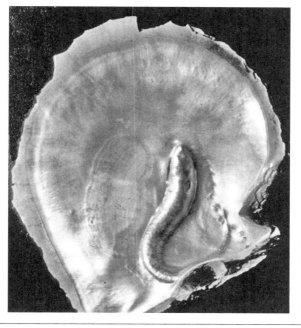

Pearlfishes (Carapidae) live inside a variety of hosts, including black-lipped pearl oysters. Apparently one pearlfish died while inside its host and was progressively covered with layers of mother-of-pearl. Courtesy of the Museum of Comparative Zoology and Harvard University

dian Oceans. Any other fish that touches the anemone's tentacles is likely to be stung by its stinging cells, then paralyzed and consumed. But resident anemonefish frequently contact the tentacles and are not stung. It is thought that the anemonefish secretes mucus that the anemone learns to recognize and does not discharge its stinging cells.

Associating with an anemone dominates all aspects of the fish's life (divers seldom if ever find anemonefish without anemones). Larval anemonefish learn the smell of their host anemone species before they leave the reef. They search for that smell when they settle out of the oceanic currents and onto a reef. Once settled, most anemonefish seldom move more than a few meters from their host, and adults remain with a single host for life. The relationship is considered mutualistic because the fish gains protection from predators on both itself and its eggs, which are laid on coral rock under the anemone. Few other fish are willing to risk being stung, so the eggs are relatively safe from predators. In turn, anemonefish chase away predatory butterflyfishes that eat anemones, remove debris from the anemone's upper surface, drop food onto the anemone, and consume anemone parasites. Also, the fish's excreted waste products may stimulate the growth of symbiotic algae that live inside the anemone. Anemones that host anemonefishes have faster growth rates, higher asexual reproduction rates, and lower mortality rates than anemones that lack a fish symbiont.

Many fishes clean other fishes (the count is up to 111 fish species in 29 fish families). These include juvenile damselfish and angelfish throughout the tropics; butterflyfish in the Sea of Cortez; surfperch in California kelpbeds; and Bluegill Sunfish in North American lakes, which is why Bluegill will nip at your moles or scabs when you go wading. However, the best-known and most highly evolved cleaning relationships involve neon gobies (*Gobiosoma*) in the Caribbean and wrasses in the genus *Labroides* in the tropical Pacific. Gobies reside at particular brain coral heads, whereas "cleaner wrasses" set up territories called cleaning stations at prominent locales in a reef complex. Species of host fish, including predators such as jacks, groupers, moray eels, and barracuda, come to cleaning stations to have parasites and dead skin removed or perhaps just to be tickled!

The wrasses advertise their presence by performing an exaggerated swimming display that involves bouncing and tail wagging. Host fish wanting to be cleaned typically assume a head-up or head-down position, hovering in the water column, blanching in color, spreading their fins, and opening their mouths. The cleaner then picks over the host's body, often entering the mouth or gill covers of herbivores and piscivores (fish eaters) alike (no one has ever found a cleanerfish in the stomach of a predator, with the exception of introduced, predatory lionfish in the Caribbean). Parasites, particularly copepods, are removed, as are mucus and pieces of tissue around wounds. Parasite numbers on hosts are lower after visiting a cleaner. Since hosts without parasites or wounds also solicit cleaning, touch alone by the cleaner must attract some fishes.

Some temperate wrasses also clean, such as the Senorita (*Oxyjulis californica*) of Mexico and California and the European Corkwing Wrasse (*Symphodus melops*). Humans take advantage of this behavior and use the Corkwing Wrasses to remove parasites from Atlantic Salmon raised in aquaculture pens, an environmentally friendlier method for killing parasites than dumping large amounts of pesticides into the water.

Cleaning is so highly evolved and predictable that two species of saber-tooth blennies (*Aspidontus*, *Plagiotremus*; Blenniidae) mimic cleaners in color and swimming behavior to gain access to host fishes. Except these aggressive mimics take a real bite out of fins and body tissue. Hosts learn to tell real cleaners apart from the mimics and the deception is most successful with young, apparently naïve, hosts.

Why do fishes form schools?

The most common reason that fishes form schools is to avoid being eaten. Several lines of evidence point to predator avoidance as the reason for schooling:

- Even fish that are solitary predators as adults, such as Largemouth Bass (*Micropterus salmoides*, Centrarchidae), stay in groups when little. Little fish have many more enemies than big fish because so many more predators can catch and eat them. When they are larger, they are too fast or too big to swallow.

- Small fish that live up in the water away from places to hide, including the so-called baitfishes such as herrings, silversides, smelts, anchovies, minnows, small tunas, and sand lances, are usually in schools. These species are called bait for a good reason.
- Fish schools are fairly disorganized most of the time, such as when the fish are moving around a reef or lake looking for food. But if a predator shows up, the fish regroup and swim closer together with coordinated movements.
- Outside of the breeding season, fish spend their time either looking for food or resting. During rest periods, their major concern is not becoming a meal for a fish that is looking for food. Fish that forage alone often form groups during their rest period. On a typical coral reef, many fish feed at night, either on the bottom or up in the water. These include grunts, sweepers, squirrelfishes, cardinalfishes, bigeyes, and many others. In a North American lake, Black Crappie (*Pomoxis nigromaculatus*, Centrarchidae) are nighttime foragers in the water column, feeding on small fishes. All of these species rest by day, often near or under some kind of cover, as part of group. Therefore, during their resting period when their major concern is not being eaten, these species occur in groups.

So fish that are vulnerable because of their size, location in the water column, closeness to cover, and kind of activity all group together and coordinate their movements when predators threaten.

Two reasons explain why being in a group decreases the chances of becoming a meal. First is the "many eyes effect." Fish in a group are on the lookout for approaching predators, and the more eyes you have looking in more directions, the more likely someone will spot the approaching shark or barracuda or pike. While everyone is watching out for predators, they are also paying attention to the behavior of the fish around them. As soon as someone on the outer edge of the group spots a predator and starts to take some sort of evasive action, all fish around it also respond. Because most predatory fish need to surprise their prey to be successful, many watchful eyes reduce the likelihood that predators can sneak up on potential victims.

The second, and ultimate, reason schooling works is because predators, especially other fishes but also squids and birds, get confused when attacking large groups of fish. Their success rate is highest when they attack small groups or single individuals. A Largemouth Bass (*Micropterus salmoides*, Centrarchidae) placed in a tank with minnow schools of different sizes is most likely to capture a minnow that is alone or in a small group.

The bass also catches the minnow sooner than when there are many fish in the school. As minnow schools increase in size, predator success goes down. When researchers watch slow-motion videos of these predator-prey interactions, they see that the predator starts to chase one prey fish, then switches to chasing another. It cannot make up its mind, which is to say it becomes confused.

Interestingly, a bass's failure rate is greatest when there are 10 or more minnows in the school, which is the same number above which people start getting confused. If someone flashes pictures of fish schools on a screen for a second and asks you to count how many fish are in the school, you are most accurate with one or a few fish. At larger school sizes, your "success" rate starts to fall off and reaches a low point at about ten or a dozen. The best explanation is that your brain, and a fish's brain, has difficulty processing information from your eyes when the number of objects exceeds ten or more. So you are no smarter than a Largemouth Bass!

Although predator avoidance is a major reason for schooling, some predators also school as a way to increase *their* hunting success. Sometimes, it is just a matter of one fish paying attention to other fish and taking advantage of another individual's feeding success. Goldfish in schools find food faster than goldfish that are alone, probably because again many eyes are better at finding food, and fish pay attention to what their schoolmates are doing. Parrotfish in schools are more successful foragers than solitary fish because schooling parrotfish are better competitors. Parrotfish often feed on the algae that grow inside the territories of damselfish. Damselfish defend these territories against all comers (including humans) by attacking and biting. Solitary parrotfish are usually driven away from the attacking damselfish. The damselfish attacks a school of parrotfish a few times but quickly gives up and the parrotfish then eat to their hearts' content.

At its most complex, some fishes hunt cooperatively, which means they help one another surround or herd fish into places where prey can be caught or maybe even confuse prey that are distracted by the presence of several dangerous predators. Blacktip sharks, pompano, sailfish, bluefin tunas, and lionfish have all been seen spreading out, taking different positions in a group, moving at different speeds, surrounding, and finally attacking prey. All this suggests cooperation among the individuals in the group, rather than each attacking prey alone and independent of the actions of other group members.

Fishes do not only school to feed or not be eaten. Fishes also form groups during the breeding season, including many species that are never found together at other times of the year. "Spawning aggregations" are known for many reef fishes, including groupers, snappers, wrasses, surgeonfishes, mullets, and parrotfishes. These aggregations occur at specific places on

Predators as well as prey form schools. These Chevron Barracuda (*Sphyraena genie*) are among the dozen or so barracuda species usually found in schools. Only the Great Barracuda (*S. barracuda*) usually occurs alone. Photo by David Hall/ seaphotos.com

the reef, year after year, and are considered "traditional." Spawning aggregations are so predictable that predators and anglers know about the times and places where they occur and take advantage of fish that are distracted by other interests. Such "spawning stupor" has been reported for a number of species, including species that are normally vary wary. In spawning aggregations, these species can be eaten by sharks or easily approached and speared. This predictable and increased vulnerability is a major reason why many grouper populations on coral reefs have been reduced to dangerously low numbers and now need protection from human hunters.

Given the extent to which schooling has been studied, you might find it interesting that fish behaviorists still argue about how to define schooling. Is it any aggregation of fishes? What should the fish be doing? How many fish make up a school? Fortunately, research with European minnows (*Phoxinus phoxinus*, Cyprinidae) has answered the last question, leaving the assignment of terms and names to the behaviorists.

Basically, one fish is a loner, two fish are a pair, and three or more fish can form a school. Few people would argue about one fish not being a school. But what about two fish? It turns out that when two fish swim together in a pair, one is usually the leader and the other the follower. But when three or more fish aggregate, each individual pays attention to what the rest of the group is doing. Basically, the group is the leader and each individual is a follower.

One other distinction can be made. Some aggregations involve fish moving at different speeds and directions, whereas other groups are highly coordinated and move as a unit. Behaviorists like to divide these groups into shoals (a loose aggregation) and schools (a coordinated group all moving in the same direction, often with fairly uniform spacing between individuals). The degree of coordination can change moment to moment. A loose shoal of feeding individuals will suddenly become a coordinated school if a predator shows up. Also, shoals of fish feeding in one area will group together as a school as they move to another area. One of the benefits of schooling while moving is that fish following close behind other fish save energy by drafting on the fish ahead, the same way that geese, bicyclists, and NASCAR drivers save energy by drafting.

Granted, these definitions are neither hard nor fast. Some schools are made up of the same individuals day in and day out, such as the French Grunt schools discussed below. In others, such as Yellow Perch (*Perca flavescens*, Percidae), the membership shifts. But again some individuals stay with the school for a long time, while others join a school passing through the area where they live but leave when the group moves outside their usual range of activity. And in some species such as Great Barracuda (*Sphyraena barracuda*, Sphyraenidae), some individuals always move around the reef as part of a group while others set up a territory that they defend.

One remaining question about schools is how fish manage to maintain the precise spacing and coordinated movements that are often stunningly shown in videos. For the longest time, behaviorists thought that spacing and coordinated turning were due to visual cues and the product of fish watching other fish carefully. Research with Pollock (*Pollachius*, a cod relative) in England has expanded our knowledge. Vision is important, but even fish that have been experimentally blinded by covering their eyes can school with other fish. So even blind fish can school, although they tend to swim a little farther apart than fish whose eyes are fully functional.

The other part of the calculation of how close or far apart to swim comes from the lateral line, the major hearing and pressure sensing organ in fishes. The lateral line is a series of pits along the side of a fish containing sensory cells that detect sound waves or water pulses, sending nervous signals to the brain via the lateral line nerve. A fish uses its lateral line to sense the water pushed by the body and tail of its swimming neighbors. If the lateral line nerve is surgically cut but the fish is not blinded, the fish will still school but it will tend to be a little closer to schoolmates than normal. Spacing in a school therefore is a balance between information coming from a fish's eyes and its lateral line. The eyes tell the fish when it is too far from a neighbor and the lateral line tells the fish when it is too close (if the fish is both blinded and has a cut lateral line nerve, it wanders off on its own).

Do fishes fight?

Fishes fight over territories, food, mates, and to decide who will be dominant and who subordinate in a pecking order. Most fights are dance-like rituals, with much fin raising, gill cover flaring, exaggerated swimming in place, circling one another rapidly, head or mouth butting, color chang-

Fishes: The Animal Answer Guide

An ichthyology student acquired this tattoo after she learned how endangered the African Coelacanth (*Latimeria chalumnae*) is.

Gars (Lepisosteidae) occur today only in North and Central America and Cuba. These primitive fishes have a bony armor and a lung and often live in swampy, backwater regions. The Alligator Gar (*Atractosteus spatula*) shown here is the largest species, growing to a length of 3 meters (10 feet) and a weight of 140 kilograms (300 pounds). Photo courtesy of Jean-François Healias, Fishing Adventures Thailand, anglingthailand.com

The Oarfish, *Regalecus glesne,* is probably the longest bony fish in the world and is responsible for many reports of sea monsters. Despite its large size, oarfish feed primarily on tiny fish and invertebrates. This individual died in the Sea of Cortez, near the shore on the eastern side of **Baja California, Mexico.** Photo courtesy of Michael Kanzler and Marcus Chua

A giant Ocean Sunfish, *Mola mola,* being cleaned by an adult Emperor Angelfish, *Pomacanthus imperator,* on a reef in Bali, Indonesia. Although molas generally live in the open sea, they often come to nearshore areas to have cleaner fishes remove their parasites. Molas are the largest (heaviest) bony fishes alive. For an idea of the size of the sunfish, Emperor Angels grow to 40 centimeters **(15 inches).** Photo by Ali Watters, www.mytb .org/Ali; used with permission

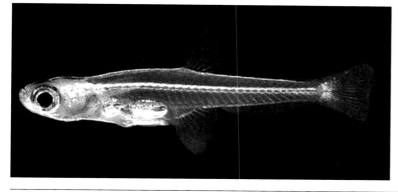

Paedocypris progenetica of Indonesia may be the world's smallest vertebrate, maturing at less than 7 to 8 millimeters (0.3 inches) length. It apparently evolved its tiny size by maturing while still retaining numerous larval traits, including being colorless, scaleless, and having poorly developed bones. From Kottelat, Britz, Hui, and Witte, 2006; used with permission

Rockfishes, such as this Canary Rockfish (*Sebastes pinniger*), are members of the scorpionfish family. Many North Pacific rockfishes live to be over 100 years old. Analysis of Rougheye Rockfish scales indicate that they may live as long as 205 years.

Barracuda are typical of burst-swimming predators with their sharp teeth. Barracuda not only impale prey but can also chop their prey up because their teeth have sharp edges, which allows them to attack large victims. Closing the mouth tightly is helped by sockets in each jaw, into which teeth in the opposite jaw fit.

The Giant Snakehead, *Channa micropeltes*, is a freshwater predator from southeast Asia. Snakeheads can grow to 1 meter (3 feet) long and over 20 kilograms (45 pounds). They live in quiet, oxygen-poor waters and survive low oxygen by being able to breathe air. Photo by Jean-François Healias, Fishing Adventures Thailand, anglingthailand.com

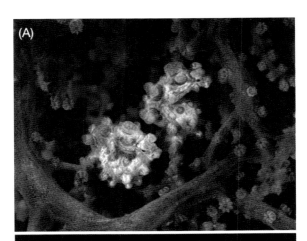

Seahorses and their relatives are masters of camouflage, resembling the physical structure in which they are most often found. (A) The Pygmy Seahorse, *Hippocampus bargibanti*, of the west tropical Pacific looks amazingly like the sea fans in which it lives. (B) The Leafy Seadragon, *Phycodurus eques*, hovers among seaweeds in South Australia. It is a protected species threatened by collecting for the aquarium trade. Photos by David Hall/seaphotos.com

The males of many fishes, in both marine and freshwater habitats, take on brilliant colors during the breeding season. This male Tangerine Darter (*Percina aurantiaca*) lives in swift-flowing waters of the southeastern United States and is one of the largest darters at a length of 18 centimeters (more than 7 inches). Photo courtesy of J. DeVivo

Many fishes have blue color in their irises. (A) A Blue-eyed Triplefin (*Notoclinops segmentatus*) from the tropical western Pacific. (B) A Golden Trout (*Oncorhynchus mykiss aquabonita*), the state freshwater fish of California. Panel A photo by Ian Skipworth; panel B courtesy of Phil Pister

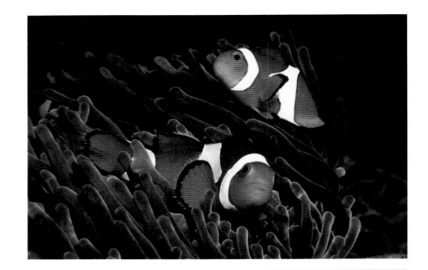

The damselfish family Pomacentridae includes many colorful,
territorial species, including the
anemone or clownfishes popular
among aquarium keepers. Photo by
Nick Hobgood

Pikeblennies inhabit worm tube burrows, and male blennies defend their tiny territories against all comers. Here a
male Bluethroat Pikeblenny (*Chaenopsis ocellata*) is attacking its image in a mirror placed on the bottom next to its
burrow. Photo by David Hall / seaphotos.com

Wrasses and other fishes pick parasites from the bodies and inside the mouths of a variety of host fishes, many of which are predators. Here a Pacific Cleaner Wrasse (*Labroides dimidiatus*) cleans inside the mouth of a fusilier (*Caesio* sp.) while other fusiliers wait their turn at a cleaning station. Photo by David Hall / seaphotos.com

How not to be seen. The large (45 centimeter, 18 inch) Pacific Spotted Scorpionfish (*Scorpaena mystes*) in this photograph is the same reddish color as much of the algae around it and pretty much invisible. It remains perfectly still on the reef, completely exposed, waiting for small fishes to pass by. The tail is in the upper right portion of the photo and the head in the lower left. Photo by Gene Helfman (who almost sat on it).

Lionfishes (*Pterois,* spp.), native to Indo-Pacific regions, are among the few fishes that are thought to be colored to inform other species of their toxic characteristics, in this case their poisonous dorsal spines. A lack of predators has contributed to the success of lionfishes as invasive introduced species. Lionfishes released by home aquarists are now common on tropical western Atlantic reefs. They feed voraciously on small reef fishes, including on parasite-picking cleaner fish that are generally immune to predation by native fish-eating species. Photo by Jens Petersen

Pacific salmon (*Oncorhynchus*) migrate from the streams of their birth along the west coast of the United States, out to the North Pacific Ocean, and return to their birth stream to spawn. Steelhead salmon (*O. mykiss*), shown here, are the same species as Rainbow Trout, but steelhead migrate to the ocean and grow much larger than rainbows, which remain landlocked.

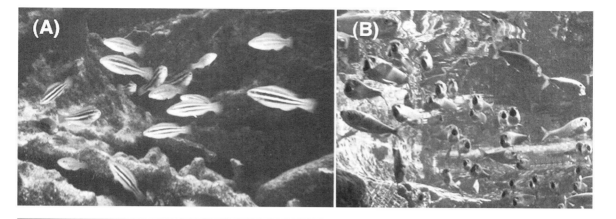

A loosely organized group of fish moving together is often referred to as an "aggregation" or "shoal," such as these foraging Striped Parrotfish (*Scarus croicensis*). (B) When most members move in the same direction with fairly uniform spacing, as in these foraging menhaden (*Brevoortia tyrannus*), the group is called a "school."

ing, and finally as a last resort, biting. All this ritualized combat is a means of deciding who will win without risking being bitten.

Even weakly electric fishes fight to establish dominance, but they do it with electricity. They do not attempt to shock one another (their electrical discharges are too weak for that). Instead they put out pulses of their electric fields that sound like buzzes and bursts of static on a loudspeaker. The winner of an electric battle keeps on discharging while the loser changes its output frequency or shuts its electric output off and departs.

Do fishes bite people?

Aside from some large sharks, very few fishes eat people. Some large fishes will bite when hooked or speared (moray eels, halibut), but those are purely defensive reactions. The same could be said for blue marlin, one of which speared a diver off Hawaii who was filming the marlin as it was being attacked by false killer whales. The diver was the slowest thing in the water and the marlin was clearly defending itself.

Bluefish (*Pomatomus saltatrix*, Pomatomidae) have been responsible for numerous attacks on bathers, especially off the coast of Florida. Bluefish get into feeding frenzies and will bite at anything in the water. Bluefish at such times are one of the few fish species that engage in "surplus killing," biting at schools of menhaden and other baitfish without actually eating them (house cats, minks, and humans are other members of the small list of animals known to engage in surplus killing). Such an aggregation of over-active predators with sharp teeth should be avoided at all costs.

The list of fish that actively and purposely attack humans is quite small, as are the fish themselves. Saber-tooth blennies in the Pacific (Blenniidae),

Fish Behavior   47

Bluefish (*Pomatomus saltatrix*) are an active, voracious predator that occur in most temperate oceans except the eastern Pacific. Bluefish schools will attack and tear apart baitfish with their relatively long and very sharp teeth. They are even known to attack swimmers. Photo from NOAA's Fisheries Collection

the same fish that mimic cleaner wrasses and bite the fins of unsuspecting hosts, are just as likely to try to take a nip out of a passing swimmer. These fish have exceedingly long teeth in their jaws, especially for a body length of only about 5 centimeters (2 inches). It is disconcerting to be snorkeling on a coral reef and feel something pinch you, only to look down and see a brightly colored, small blenny swimming back to the safety of the reef.

The only other fish known to actively attack humans for food is a 25 centimeter (10 inch) cichlid, *Docimodus johnstoni*, in Lake Malawi, Africa. This fish is a fin-biter, which means it attacks other fishes and chomps off pieces of fin. The native name for the cichlid is *chiluma*, which means "the thing that bites." Apparently, the fish is not particular about what it chomps.

What about piranhas (*Serrasalmus*, *Pygocentrus*; Characidae)? It turns out that most instances of people being "eaten" by piranhas had actually drowned first; the fish scavenged on the remains. If we consider this behavior as "predation," then we would have to include a long list of scavengers, including hagfishes, as predatory on people. A growing number of piranha attacks have occurred recently in Amazonian reservoirs, but almost all involve waders who wander near a piranha nest and are bitten once, usually on the foot or leg, by a nest-guarding adult. At this level, a damselfish guarding its territory can be considered just as aggressive, although less well armed.

How smart are fishes?

How smart does a fish have to be? Fishes are very good at doing the things they have to do. They probably would not do well on math tests or in a spelling bee, but those are not tasks that evolution has prepared them to complete. But when it comes to getting around a reef, finding food or a mate, avoiding predators, or migrating to a place and back again, fish show what we would consider common sense and good learning ability.

We have many everyday examples of learning in fishes. Every angler knows that the biggest fish are often the hardest to catch, probably because

Fishes: The Animal Answer Guide

they've been caught before and they're not going to fall for that trick again: fool me once, shame on you; fool me twice, shame on me. If you have ever owned a goldfish, you have seen it swim up to the front of its bowl when you come into the room because it has learned that you might feed it. Scuba divers go to particular places to feed particular individual, tame moray eels, wolf eels, groupers, Napoleon Wrasse, barracuda, and even sharks. These fish approach divers looking for handouts. Where fish have not been fed, they are more likely to be cautious. Where they have been fed, they have learned that divers are—or more correctly will offer them—an easy meal.

Learning has been shown in many fishes in many other situations. For example, young French Grunts (*Haemulon flavolineatum*, Haemulidae) spend their daylight hours in resting schools above coral heads or among sea urchin spines. At dusk, they migrate into nearby grassbeds, the schools break up, and the fish feed as individuals on small animals. At dawn, they migrate in the same group back to the same daytime resting spot, using the exact same routes they used in the evening, only in reverse. You can set your watch by their timing, both evening and morning. And you could draw a map of the reef and grassbed and plot these routes exactly to within a few feet or less as if you were drawing a road map.

So how is it that a young grunt that settles on a reef and joins a school knows how to get between reef and grassbed each evening and morning? Every grunt could have an instinctual map of every reef, hardwired into its brain. But when these fish spawn, their floating larvae can be carried many miles away from home. The larvae could land on any one of a million places. That's a lot of maps in an admittedly small brain. Instead, it has been shown that new arrivals learn the route by following the residents already there. And they are quick learners because it only takes them a couple of evenings and mornings to perform the migration correctly, even if their teachers are then taken away.

Fishes are also very good at learning who their competitors and predators are and are not. One study of Beau Gregory Damselfish in the Virgin Islands found that male damselfish could tell apart as many as 50 different species of other reef fishes. It had a checklist of bad actors. Fishes that did not eat algae were allowed to pass through the territory undisturbed. Competitors that were likely to eat the algae in the territory were driven off, and large competitors (such as the parrotfish mentioned above) were driven off more energetically than smaller algae eaters. The most vigorous attacks were directed at trespassers that were likely to eat the eggs that the male damselfish was guarding. The territorial male responded differently depending on the value of the resource it was guarding and the threat a trespasser presented.

Fish Behavior

Juvenile grunts (Haemulidae) in the Caribbean hover during the daytime in the protection of coral fingers or long-spined sea urchins. In the evening, they migrate to nearby grassbeds to feed.

In a similar manner, damselfish recognize how dangerous different potential predators are and then respond accordingly. When models of trumpetfish (*Aulostomus*), a common predator on small reef fishes such as damselfish, were passed over Threespot and Bicolor damselfish, the damselfish responded according to the size and "behavior" of the model. Trumpetfish that were too small to eat the damselfish were largely ignored. Large trumpetfish caused the damselfish to stop and look up. If a large trumpetfish model stopped and pointed head down (the usual pre-attack behavior), the damselfish dove into its refuge hole. Small damselfish responded more strongly to both small and large trumpetfish, which makes sense given that small damselfish are a potential meal for a larger range of predators. Damselfish responded in a graded manner to a range of predators pretty much as you would expect based on how dangerous the predator was.

All this makes sense when you think of what a damselfish has to do during the day. It is feeding, guarding eggs, courting females, and chasing away competing males. At the same time, hundreds of fish of dozens of different species are swimming around it every hour of the day. If every time a predator swam by, the damselfish dove into its hole—and there are dozens of predator species on the reef—the damselfish would lose valuable time that could be spent feeding, guarding eggs, courting females, and so on. A fish needs to be able to assess the behavior of potential predators and take an action that is appropriate for that particular situation. It just makes good common, and evolutionary, sense.

Fishes also learn the features of their area and rely on their memory to find their way around. They keep a map in their brain. This is known for butterflyfishes on reefs and for Mexican Blind Cavefish (*Astyanax mexicanus*, Characidae) in aquariums. This totally blind species swims actively

Fishes: The Animal Answer Guide

The Mexican Blind Cavefish (*Astyanax mexicanus*, Characidae) is a totally blind species that develops eyes when raised in the light but is eyeless when raised in darkness. Photo by Ltshears

but never bumps into objects in the tank. As it swims, it pushes water ahead of it, as does any object moving through water. The cavefish detects objects by sensing when the water it pushes bounces back from an object. After a fish has been in a tank for a while, it memorizes where various things are located—plants, rocks, filters, plastic treasure chests—using its memorized map to guide it. If you then move things around in the tank, and this is an experiment that anyone can do at home, the fish will bump into objects because they were no longer where they were supposed to be.

Do fishes play?

Sadly, most fishes fall down as playmates. There is very little evidence to suggest that fishes engage in what we recognize as play, such as manipulating objects in a tank for no apparent reason. One of the few examples of play in fishes comes from what may be one of the more intelligent fishes, an African species known as the elephantfish (*Gnathonemus petersi*). This family (Mormyridae) is well studied because its species are weakly electric (see "What are electric fishes?" in chapter 2). Individuals create a weak electric field around their bodies and then detect disturbances to the electric field that are strong or weak conductors of electricity. A weak conductor would be a rock or stick; a strong conductor would be another animal because animals have ions in their blood that make their blood a good conductor. Elephantfish can apparently distinguish different species and sizes of objects and their direction and speed of movement based on disturbances to their own electric field.

An elephantfish processes electric information in the cerebellum of its brain, and elephantfish have the largest cerebellum for their body size of any known fish. And elephantfish have been observed playing. They will take a small ball of aluminum foil—a very good conductor of electricity—and carry it to the outflow tube of the aquarium filter such that the ball is

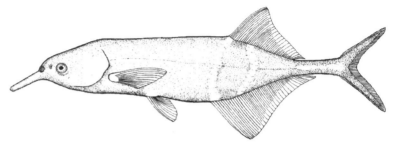

African elephantfishes produce a weak electric field to gain information about their surroundings. Peters Elephantfish (*Gnathonemus petersii*) is the only fish that has been observed playing with objects. From Helfman et al. 2009; used with permission of Wiley-Blackwell

pushed across the tank by the water currents. The fish will do this repeatedly, for no apparent reason. This meets the requirements of most definitions of play, which is pretty hard to define. In mammals, our large cerebrum directs our play, so why not use the extra "mental capacity" of the elephantfish's oversized cerebellum for the same reason. That is the best explanation anyone has come up with.

Needlefishes (Belonidae) have been observed jumping over floating sticks or pieces of seagrasses. Perhaps they are trying to remove parasites or maybe they are really playing.

Do fishes talk?

Some talk more than others, but the names of many fishes—grunts, grunters, trumpeters, croakers, drums, drummers, sea robins—are a good indication of how widespread sound production is in fishes. The sounds produced are variously described as grunts, pops, thumps, squeaks, squeals, foghorns, boat whistles, groans, booms, rumbles, purrs, knocks, clicks, hums, and buzzes. But these words apply to the fishes we can hear easily, usually when captured. Our ability to detect fish sounds has improved greatly with the development of new and more sensitive underwater microphones, which means the list is always growing.

The underwater world is anything but silent. Jacques Cousteau's book *The Silent World* was inaccurately named, perhaps because it is difficult to hear fishes underwater when one is diving with noisy, bubble-producing scuba. Fishes vocalize to attract mates, when defending nests or territories, when attacked by predators, to keep schools together, and to warn their young of danger. They make sounds using a variety of body parts, including drumming with the gas bladder via a special muscle called the "sonic muscle"; rubbing fin spines, fin bases, and skull and jaw bones together; and clicking jaw teeth or grinding pharyngeal (throat) teeth together. In many, a sound is produced at one place (such as jaw bones) and then amplified (increased) by traveling through and vibrating the gas bladder.

Fishes: The Animal Answer Guide

Many fishes emit distress sounds when grabbed, prodded, poked, or even surprised (e.g. herrings, characins, catfishes of many families, cods, squirrelfishes, sea robins, grunts, drums, porcupinefishes, triggerfishes). At least three families (cods, squirrelfishes, groupers) produce distinctive sounds when merely approached by predators, and when squirrelfishes emit their staccato predator-awareness sound, other squirrelfishes dive for cover. Schooling is considered an important predator avoidance tactic, and some schools stay together by producing sounds. Pacific and Atlantic herring (*Clupea*, Clupeidae) school at night and keep their schools together by emitting "FRTs" (fast, repetitive tick-like sounds). The FRT is created by releasing small bubbles from their anus for as long as 7 seconds. Try that at home!

Sound is a regular part of the courtship and spawning of many fishes. Male sturgeons, minnows, characins, codfishes, toadfishes, midshipmen, sunfishes, grunts, drums, seabasses, darters, damselfishes, cichlids, blennies, and gobies vocalize to attract females during courtship. Males often perform elaborate visual displays while vocalizing by swimming rapidly in exaggerated patterns, erecting fins, and jumping out of the water. Males often call faster as a female gets closer. In some fishes, such as African cichlids, the male's vocalizations stimulate egg production in females, similar to what happens in seasonally breeding birds.

Although male calling is the general rule, in a few species both sexes call during a spawning bout. This exchange takes a twist in hamlets (*Hypoplectrus*), small members of the seabass family Serranidae. These fish are hermaphroditic, individuals producing both sperm and eggs. During spawning, they take turns being male and female, the "male" emitting a courtship call and the "female" a spawning call. As individuals switch roles during a spawning bout, they also switch the sounds they produce.

Toadfishes and the related midshipmen (*Opsanus*, *Porichthys*; Batrachoididae) are among the loudest and noisiest of fishes. The humming of midshipmen in San Francisco Bay is so loud during the breeding season that it keeps people living on houseboats awake at night. The Oyster Toadfish (*Opsanus tau*) produces a boat whistle call to attract females by vibrating its swim bladder. The muscles involved contract at a rate of 200 Hz (200 times per second). This is the fastest contracting vertebrate muscle known, even faster than the shaker muscles at the base of the tail of rattlesnakes.

How do fishes avoid predators?

As mentioned earlier, outside of the breeding season, most of a fish's time is spent either eating or avoiding being eaten. And although eating is obviously important for survival, many fish can go for long periods without eating and still survive. But one encounter with a predator can ruin your whole day. Hence the importance of avoiding predators is obvious in almost all aspects of fish anatomy, ecology, and behavior and deserves some detailed exploration.

Animal behaviorists describe predatory behavior as a sequence or cycle of events, one following the other, that must be more-or-less followed for a predator to be successful. This sequence includes searching for and detecting prey, pursuing prey, attacking and capturing, and finally handling. Predator avoidance not surprisingly involves stopping the sequence as soon as possible. A prey animal that cannot be found does not have to worry about running away or being difficult to swallow. Fishes show a tremendous variety of intriguing and fascinating adaptations to thwart predators at each point in the cycle.

By Avoiding Detection. The primary means a fish has of not being found is to not be seen. The easiest solution here is to hide in a hole, among vegetation, or dive into the sand. Many small reef fishes take refuge in the reef when frightened, including damselfishes, wrasses, blennies, and most fishes that are night-active (squirrelfishes, cardinalfishes). Many lake species, such as sunfishes, swim quickly into weedbeds when frightened. Kelpbeds are a good place to hide for surfperches, grassbeds for wrasses and rabbitfishes. Sand diving is actually pretty remarkable when you consider that compact sand is fairly solid, but razorfishes and sandlances among others do it with ease.

However, hiding out of sight has some disadvantages, chief among which is that you cannot do much else while you are hiding. Also, it is hard to peek out to see if danger has passed without exposing yourself. One way around this is to hide most of your body but leave your eyes exposed so you can still be aware of predators or potential prey. Many fishes bury themselves in the sand or mud with just their eyes showing (angel sharks, stingrays, lizardfish, sandfish, weeverfish, stargazers, and flatfishes).

Most fishes however remain out in the open when frightened. Larger fish can of course swim away, but most predators are larger than their prey. All other things being equal, larger fish can swim faster than smaller fish.

Rather than trying to run away, the most common defense is to be part of a group. Schooling was discussed earlier and occurs in all habitats by all types of fishes (but especially among silvery fishes; see "Why are so many fishes silver?" in chapter 3). Because of the many-eyes and confusion effects, fish find safety among a large number of schoolmates.

Perhaps the most common way to not be seen is to be colored to match your natural surroundings, to be camouflaged. Fish do this in two ways, by either appearing like the background or by disappearing altogether. Background matching, often called mimicry, involves having coloration that

Garden eels are small (to 50 centimeter, 20 inch) members of the conger eel family Congridae that feed on zooplankton but live in the bottom, an unusual combination. When not frightened, they emerge most of the way from their burrows, face into the current, and pick off passing zooplankton. When threatened, they back down into their burrows and disappear, the entire group descending together. This group has just emerged and the heads of some individuals are barely visible.

matches both the actual colors and the color pattern of the background. Flatfishes (order Pleuronectiformes) are masters of this, changing the color of their upper side (actually one of their sides since they lie on the other side, not their belly) to match the bottom. This usually just means taking on the general coloration of the sand or mud around them. But some flatfishes are so good at this that one placed on a checkerboard will actually take on a checkerboard coloration.

Often, rather than trying to match the background, fish live in a habitat that is the same color as their skin. Many scorpionfishes, toadfishes, and crocodile fishes on reefs and sculpins, darters, and suckers in streams have a blotchy color pattern that blends perfectly with the algae-covered rocks where they most often live. Many a diver has had a close call placing his or her hand down on the bottom only to have a scorpionfish with its poisonous spines dart away where their hand was about to be placed.

Many green colored fishes live among plant stems or green algae, including wrasses, pipefishes, gunnels, and small sculpins. Shrimpfish (Centriscidae) hover head down amongst the spines of long-spined sea urchins. Their camouflage is improved by being long, narrow, and light-colored with a thin black stripe that runs the length of the body. Some fish have fleshy appendages that increase their resemblance to the algae where they live, as in sculpins, sargassumfish, and the spectacular Australian Leafy and Weedy Seadragons (Syngnathidae).

The neatest trick of all is becoming invisible while in plain view up in the water. Fishes achieve invisibility via their coloration, and they do this in one of two ways, both of which rely on how light penetrates the surface and is distributed in water (see "Is there a reason for the color patterns of fishes?" in chapter 3). The most common coloration among fishes is countershading (dark on top grading to a light colored belly). A countershaded fish blends in against the open water background no matter which angle you look at. Fishes also "disappear" by having silvery sides that reflect light of the same strength and color as the background, thus again blending into the background (for an explanation of how silvery fishes disappear, see "Why are so many fishes silver?" in chapter 3).

BY AVOIDING CAPTURE. All the emphasis on hiding and camouflage in fish anatomy and color makes sense because a fish that is not found by a predator does not have to worry about escaping capture. But many fish are easily seen. They discourage predatory attacks by being difficult to capture. Speed is one defense but, as mentioned before, little fish have a hard time swimming faster than bigger (predatory) fish. A smaller fish can turn faster than a larger one, so a couple of quick turns might buy time if the turns are followed by diving into a hole. But ultimately relying on speed is a risky tactic for a small fish trying to elude a large predator.

One way that small fishes gain a speed advantage is by leaving the water altogether, by jumping, leaping, and finally, flying. Many fish jump out of the water when pursued, including minnows, Asian carps, halfbeaks, needlefishes, sauries, killifishes (topminnows), silversides, Bluefish, mullet, and mackerel. Most of these fish already live near the water's surface, so it is a short trip into the air. As soon as they leave the water, their speed increases, often doubling. If the predator does not also leap out of the water, the prey can rapidly outdistance a pursuing predator.

Flyingfishes (Exocoetidae) are the unquestionable masters of this tactic. Flyingfishes are small, water column swimmers and, not surprisingly, are silvery, but when detected and chased by a predator, they take flight. Actually they glide, and most parts of their bodies are clearly adapted to maximize the time they can spend above the water surface. Flyingfishes come in two major varieties, two-wing and four-wing. In all, the pectoral fins are enlarged for gliding (four-winged flyingfish also have enlarged pelvic fins). The front edges of the pectorals are stiffened, and the fin curved downward, like an airplane wing. The lower part of the tail fin is longer than the upper part. As the fish breaks through the water surface and picks up speed, it spreads its pectoral fins and glides downwind, often on a curving path. When it slows and loses altitude (some can get as high as 7.5 meters, or 25 feet, above the water), it touches the water surface first with the longer lower lobe of the tail fin, rapidly moving the tail back and forth. It therefore "swims" with just that part of the fin and regains speed and takes off on another downwind glide. In this manner, a flying fish can undertake 12 touch-and-go segments, cover over 400 yards, achieve speeds of 50 miles an hour, and remain airborne for 30 seconds. During all of this, the flying-fish is probably no longer visible to the pursuing predator, and its eventual landing locale is difficult to predict because of the curving path.

Other fishes glide and some even propel themselves out of the water. Best known are the South American freshwater hatchetfishes (Gasteropelecidae) that also have greatly enlarged pectoral fins. Hatchetfishes vibrate their pectoral fins using pectoral muscles that may account for 25% of their body weight to launch themselves out of the water (see "Can any fishes fly?" in chapter 2).

A common means of discouraging pursuing predators is to have poisonous skin or spines and to advertise the fact. Warning coloration—bold, contrasting blacks, whites, and yellows—is best known in wasps, poison dart frogs, and skunks. In fishes, advertisement occurs via slow movement and erection of spines. Warning coloration is uncommon in fishes, even among those with toxic spines. The marine eel catfishes (*Plotosus*, Plotosidae) of the Indo-Pacific have highly venomous spines and bold white lines on their dark bodies. They are among the few catfishes that are active during the day, occurring in tightly grouped schools that swim across a reef, feeding actively. Their conspicuous color and behavior are easily interpreted as "stay away" signals to any would-be predator (they are still eaten, by sea snakes, themselves poisonous).

Other fishes with possible warning colors include the slow moving lionfishes (*Pterois*, Scorpaenidae), with long, flowing, red, black, and white poisonous spines. Their bold behavior is characteristic of animals that have little to fear from predators and may represent advertisement of their toxic trait. Some gobies (*Gobiodon*, Gobiidae) secrete skin toxins that cause loss of equilibrium and even death in predators. The most toxic gobies are also the most brightly colored, suggestive of warning coloration. An alternative explanation is that by being highly toxic, these fish can afford to have bright colors for interacting with other gobies without having to worry about being conspicuous (both advertisement and behavioral interaction could occur). Some surgeonfishes (Acanthuridae) have a bright orange patch around the sharp, toxic spines at the base of their tail, which again could be a warning sign.

Spines do not have to be toxic to be a good defense. Stiff hard spines can cause painful injuries. When a fish erects its spines—as many do when confronted by a predator—it increases its body depth, making it more difficult to be swallowed. A Bluegill Sunfish (*Lepomis macrochirus*, Centrarchidae) increases its body depth by about 40% by erecting its fins, making it a larger and hence less desirable food item for most predators. Other fishes

The Striped Eeltail Catfish (*Plotosus lineatus*, Plotosidae) is a fish that advertises its toxic spines. Most catfishes are solitary and only active at night, but Striped Eeltail Catfish move actively around Indo-Pacific coral reefs in large schools during the day. Their bold striped color probably warns potential predators about their spines. These catfish are being followed by small jacks (Carangidae) that will feed on animals disturbed by the catfish.

Photo by David Hall/seaphotos.com

such as triggerfishes may dive into a hole and then erect and lock their spines, making them difficult to dislodge.

The ultimate spine erection occurs in the porcupinefishes (Diodontidae). Puffer, or balloon, fish have modified jaw muscles and bones, skin, stomach, scales, body musculature, and a body cavity all involved in a defensive tactic whereby a threatened fish turns itself into a large, inedible, spiny sphere. The porcupinefish fills its stomach with water, increasing its volume by 50 to 100 times. In the process, the skin stretches, causing specialized spines in the skin to erect and interlock, increasing body volume threefold and creating an object too large and too difficult for most predators to swallow.

By Avoiding Being Handled. Large size, erect spines, and toxic skin all make handling and processing a prey individual more difficult. To these can be added anatomical characteristics that make dismembering or swallowing more difficult. Most predatory fishes are gape-limited, meaning they have to swallow their prey whole. Most predators lack the cutting teeth of specialists such as piranha, Tigerfish, Bluefish, barracuda, and sharks and cannot eat what they cannot swallow whole. Importantly, a miscalculation of prey body size by a gape-limited predator can be deadly if the prey gets stuck in its throat.

So anything a prey individual can do to make it harder to hold, dismember, or swallow will be favored. Slime production makes holding a prey more difficult, which is one hypothesized function of the abundant slime produced by hagfishes and freshwater eels. Distasteful substances in the skin, as in soapfishes (*Rypticus*, Serranidae), marine catfishes (Ariidae), toadfishes (Batrachoididae), and clingfishes (Gobiesocidae), may cause a predator to spit out a fish it had begun to swallow. And a physical defense against handling,

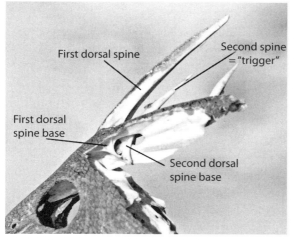

First dorsal spine

Second spine = "trigger"

First dorsal spine base

Second dorsal spine base

Triggerfishes (Balistidae) lock their dorsal spines in the erect position, making the animals very difficult for predators to swallow. The fish (or a person) unlocks the spines by pushing down on the second, smaller, dorsal spine, which is the trigger. This causes the base of the second spine to push against the base of the first spine, which depresses the first spine.

one that goes back more than 500 million years when some of the very first fishes appeared, is body armor. Thick scales, scutes, bony exteriors, and thick or hardened skin are all lines of defense against biting and swallowing predators. Some of the most primitive bony fishes—including sturgeons, bichirs, gars, Bowfin, Arapaima—as well as some evolutionarily advanced fishes—armored catfishes, boxfishes, triggerfishes, ocean sunfishes— possess dermal armor or very thick skin that protects them from predators.

Some fishes have one more response to being handled. Although this defense works very late in the predation cycle, some might say too late, it is widespread enough to suggest that it is a successful anti-predator defense. These fishes produce a chemical called an alarm substance that is only released if the skin of the prey is broken, such as when a fish is bitten by a predator. Alarm substances are best known in the ostariophysan fishes (minnows, suckers, catfishes, loaches) but have evolved repeatedly in other groups including salmonids, livebearers, sculpins, darters, Yellow Perch, cichlids, and gobies, and perhaps in galaxiids, killifishes, and silversides. Common to most of these groups is schooling behavior, and it is the reaction of schoolmates that has been best studied. When the alarm chemical is released into the water, the injured individual's schoolmates school tightly and move away from the area where alarm substance is released. In addition, many predators are attracted to the area where the substance occurs, which is another reason why the prey school should leave the area.

Of what possible advantage is it to an injured fish to have its schoolmates abandon it at such a critical moment? Perhaps there is an evolutionary advantage because its schoolmates may very well be brothers and sisters who are saved by such sacrificial behavior.

But the response of other minnows in the area may not be the reason that alarm substances evolved. Alarm substances can be grouped with another reaction of many fishes when grabbed by a predator, namely they scream. Many fishes emit distress sounds when held, prodded, or speared, as discussed earlier. These cries are often called "release calls" and happen in many other animals, including frogs, birds, baby alligators, and numerous mammals, ourselves included. Such a distress call might be an advertisement of a feeding opportunity meant for an even larger predator. A small predator with a fish in its mouth, one it is having difficulty swallowing, is handicapped. It cannot swim as fast, turn as fast, dive into a hole, or do many of the things that fishes do when avoiding predators. Hence the opportunity provided by capturing prey carries the risk of turning into prey yourself. Perhaps a chemical alarm substance released into the water, combined with a scream, will attract another, larger predator. Rather than risk becoming a meal, a small predator upon detecting the alarm substance might release the prey fish it held in its mouth.

Chapter 5

Fish Ecology

Do fishes migrate?

Many fishes are very mobile, undertaking daily, seasonal, reproductive, and life cycle migrations. Daily migrations can be measured in meters or feet, whereas annual and life cycle migrations can crisscross entire oceans.

Fishes in both fresh and sea water move back and forth between habitats on a daily basis. In most locales, fishes feed during either the day or night. When not feeding, fishes rest, usually in a place safe from predators. This change in activity often requires a change in habitat because the best places to feed, such as up in the water or over a sandy area, are not good places to hide. Fishes usually migrate between feeding and resting locales at dawn and dusk. Migrations may involve movement of a hundred meters (325 feet) or less between a reef and nearby grassbed or just a few meters for fishes that hide in the reef but feed in the water directly above it. These movements, at very predictable times and places, provide feeding opportunities for another ecological group, the predatory fishes. Predators do not migrate but instead position themselves along the migrators' corridors, picking off individuals that move too early or too late or depart from the safety of the migrating school.

On coral reefs, damselfishes, surgeonfishes, parrotfishes, herrings, drums, squirrelfishes, and grunts make daily feeding migrations. In kelpbeds, silversides, seabasses, surfperches, damselfishes, wrasses, and croakers move between daytime and nighttime habitats. In temperate freshwater habitats, freshwater eels, catfishes, minnows, and yellow perch move between habitats at dusk and dawn. At a larger scale, scalloped hammerhead sharks (*Sphyrna*, Sphyrnidae) make daily movements between daytime locales above seamounts and nighttime foraging areas in open water, several kilometers (miles) away.

One other type of daily migration that occurs at small and large scales is the vertical migration of fishes that live in the water column. In the ocean, dozens of species move at dusk from deep, cooler waters to feed in shallow, warmer waters at night. Some of these fishes, such as lanternfishes, lampfishes, bristlemouths, lightfishes, and cookie cutter sharks, travel as much as 500 meters (1600 feet) upward in the evening. On coral reefs, damselfishes, butterflyfishes, small anthiine seabasses, and triggerfishes feed up in the water by day and are replaced at night by cardinalfishes, squirrelfishes, and copper sweepers. The nighttime feeders hide by day inside coral heads, where they are replaced in the evening by the daytime feeders. In temperate lakes, alewife (*Alosa pseudoharengus*, Clupeidae) migrate upward in the evening to feed on small mysid shrimp that also move up and down. Coelacanths (*Latimeria*, Latimeriidae) are among the few fishes that move down to feed at night.

Seasonal, annual, and reproductive migrations occur in the ocean and in large rivers. These migrations often involve movement among the different habitats that fishes occupy at different times in their lives. Many reef fishes migrate several kilometers (miles) to spawn at traditional locales, including seabasses, snappers, wrasses, parrotfishes, and surgeonfishes. More than a hundred thousand Nassau Grouper (*Epinephelus striatus*, Serranidae) moved from as far away as 110 kilometers (68 miles) to spawn at one site in the Bahama Islands, until their predictability allowed them to be fished out.

Large catfishes in the Amazon (South America) and Mekong (Southeast Asia) rivers move hundreds of kilometers (miles) upstream during annual spawning migrations. Their young move downstream with currents to nursery habitats. For many of these and other tropical river species (osteoglossid Arapaima, mormyrid elephantfishes, large characins, minnows, and gymnotid knifefishes), adults migrate up tributaries and onto floodplains to spawn. Colorado Pikeminnow (*Ptychocheilus lucius*, Cyprinidae) make similar long-distance migrations in the Grand Canyon, as do sturgeons and paddlefishes in the Yangtze River of China (or at least they did before pollution, overfishing, and dams exterminated them). Atlantic Herring (*Clupea harengus*, Clupeidae) spawn off southern Norway and then move to feeding grounds around Iceland, a distance of about 1,700 kilometers (1,000 miles). The young develop in northern Norway before moving to the feeding grounds.

Many open ocean fishes are called "highly migratory species" because their movements carry them around and through entire ocean basins. Included are tunas, billfishes, and some sharks. The Blue Shark makes return trips between North America and Europe, a distance that exceeds 16,000

The American Eel (*Anguilla rostrata*) migrates at dusk and dawn. These eels are part of a large population that lives in a dark, limestone cavern in Northern Florida. By day, the eels hide among the holes in the limestone and then migrate from the cavern each evening to forage in shallow surface waters through the night.

kilometers (9,900 miles), and White Sharks are now known to migrate between central California and Hawaii, a minimum one-way distance of 3,800 kilometers (2,350 miles), and between South Africa and western Australia (a roundtrip of 22,000 kilometers, or 13,600 miles). Whale Sharks move around and across oceans, arriving at particular places to feed on eggs spawned by reef fishes and invertebrates. A Whale Shark tagged off Mexico traveled 13,000 kilometers (8,000 miles) across the North Pacific Ocean. Another tagged in the Philippines traveled to Vietnam 4,567 kilometers (2,830 miles) in only 2.5 months, while another tagged off Malaysia moved 8,025 kilometers (4,975 miles) over a 4-month period. These movements are a major cause of population declines among oceanic species, because fishing for these species is largely unregulated once they leave the territorial waters of any one nation.

The ultimate lifetime migrations are undertaken by salmon in the North Pacific and North Atlantic oceans. All spawn in fresh water, sometimes hundreds of kilometers (miles) from the ocean, and then move downstream to the sea as juveniles. Depending on species they may spend 1 to 6 years at sea before returning to the stream where they were born, where they spawn and most die. While at sea, some make annual migrations around entire ocean basins such as the North Pacific. A large Chinook Salmon (*Oncorhynchus tshawytscha*, Salmonidae) could theoretically swim minimally 7,000 kilometers (4,300 miles) just moving from the Snake River to the Aleutian Islands and back to the Snake, not counting the area covered during the 5 or so years it spent feeding in the North Pacific. Salmons locate their birth stream by following its scent, a scent they learned on their way out to the ocean as juveniles. How fishes such as tunas and salmons find their way around ocean basins remains something of a mystery. However, these fishes have small particles of magnetic material in their brains that

The Mekong Giant Catfish (*Pangasianodon gigas*, Pangasiidae) is probably the world's largest catfish, growing to 3 meters (10 feet) long and weighing more than 300 kilograms (660 pounds). Unfortunately, few fish of this size exist today. These and other large, endangered catfishes undertake annual spawning migrations through entire river basins and are threatened by overfishing and dams that block their migrations. Photo courtesy of Zeb Hogan, University of Nevada–Reno

may serve in some way as a compass. Tunas have a translucent "window" on top of their heads above their brain that is light-sensitive and may also be involved in long-distance migrations.

How many fish species live in rivers versus lakes?

Of the 31,000 or so species of fishes, about 41% (11,500) live in fresh water. Another 1% (280) spend at least part of their lives in fresh water. About 99% of the world's fresh water occurs in lakes, so we would expect more fishes in lakes. However, many lakes are too deep or too unproductive for fishes because they are covered with ice for much of the year. Our best guess is that about 15% (1,700) of the 10,250 described freshwater fish species are restricted to lakes, meaning that 85% of freshwater fishes live in rivers and streams (fewer than 1% live in caves and springs). In North America, Europe, and Asia, most fishes live in southern regions. These areas have few large lakes, so again rivers are more species-rich. Also, the tropical regions of Asia, Africa, and South and Central America contain the largest number of species. And again, fish diversity is greatest in tropical rivers such as the Mekong, Zaire-Congo, Amazon, Orinoco, Uruguay, and Parana-Paraguay.

Although the world's rivers have been factories of fish evolution, lakes have also been important. The great lakes of Africa (Victoria, Tanganyika, and Malawi) alone contain between 1,500 and 2,000 species, mostly cichlids. The five Great Lakes of North America (Huron, Ontario, Michigan, Erie, and Superior), in contrast, contain "only" 235 fish species.

Fishes: The Animal Answer Guide

How many fish species live in the ocean?

Approximately 58% (16,000+) of fish species live in the ocean, but some regions of the sea house more species than others. Nearshore areas from the coastline down to about 200 meters (650 feet) are richest in diversity, with about 12,600 species (cods, seabasses, croakers, smelts, surfperches, flatfishes, all coral reef families). Fishes that live near the bottom in water deeper than 200 meters are next with about 1,800 species (grenadiers, rattails, snailfishes, cusk eels, and tripodfishes). Open water fishes that swim above the bottom below 200 meters include perhaps 1,400 species (deep-sea anglerfishes, lanternfishes, hatchetfishes, whalefishes, dragonfishes), whereas fishes that swim in the open sea at depths of less than 200 meters make up the smallest group of about 360 species (tunas, billfishes, mahi-mahis, flyingfishes, ocean sunfishes).

How far down in the ocean do fishes live?

The deepest living fish is a cusk eel, *Abyssobrotula galatheae* (Ophidiidae) that was caught in a bottom trawl at 8,370 meters (27,500 feet) in the Puerto Rico Trench. At such depths, fish would experience pressures of 800 atmospheres, or approximately 12,000 pounds per square inch. Snailfishes (Liparidae) have been filmed in deep-sea trenches at depths of 7,560 meters (24,800 feet) and 7,700 meters (25,300 feet). Sharks do not occur as deep as bony fishes. Most sharks live above 3,000 meters (9,840 feet), the few exceptions include sightings of a Portuguese shark (*Centroscymnus*, Somniosidae), at 3,690 meters (12,100 feet) and an unidentified dogfish (Squalidae) at 4,050 meters (13,280 feet).

Which geographic regions have the most species of fishes?

Geographically, the highest diversities are found in the tropics. The Indo-West Pacific area that includes the western Pacific and Indian oceans and the Red Sea has the highest diversity for a marine area (3,000 species). Next highest is the western Atlantic region that includes the east coast of the United States, the Gulf of Mexico, and the Caribbean. It is home to at least 1,200 fish species. The eastern Pacific region from California to the coast of Peru houses perhaps the same diversity. The tropical eastern Atlantic, including the Mediterranean Sea, contains fewer species, totaling only about 550 species of shore fishes.

Tropical areas again contain the most freshwater fishes, although many locales remain to be explored and their fishes described. South America

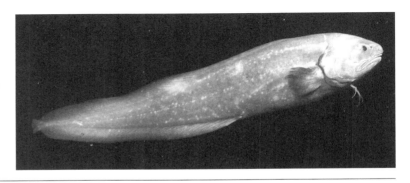

The deepest living fishes are in the cusk eel family Ophidiidae, subfamily Neobythitinae. This neobythitine, the Giant Cusk Eel (*Spectrunculus grandis*), occurs as deep as 4,800 meters (15,750 feet). It is large for a deep-sea fish, growing to 1.3 meters (4.25 feet). Photo by NOAA/MBARI

is home to perhaps 6,000 freshwater fishes, whereas Africa and Southeast Asia both house maybe 3,000 species, but the numbers grow continually with new discoveries.

Are there fishes in the desert?

Although deserts seem unlikely places for fishes, a small number of fishes do survive in deserts around the world. The most likely place to find desert fishes is where springs emerge and form ponds and streams. Because such springs are often few and far between, the fishes that live in them occur nowhere else in the world. Species of desert pupfishes (Cyprinodontidae) in North America are often found in only one spring, in such unwelcoming (for fish) locales as Death Valley, Ash Meadows, Salt Creek, and Devil's Hole. The endangered Devils Hole Pupfish (*Cyprinodon diabolis*) occurs in a single spring in Nevada, its entire habitat restricted to a limestone shelf that measures only 18 square meters (about 200 square feet) in area. It probably has the smallest range of any fish and perhaps of any vertebrate.

For desert fishes that do not have permanently available spring water, disappearing water is one of many challenges. As water dries up, temperatures rise, salts are concentrated, oxygen levels drops, carbon dioxide increases, and competition and predation intensify. Desert fishes adapt to these conditions in several ways. Some hatch from eggs during brief rainy seasons, grow quickly, and spawn and die before the next dry season. Their eggs survive in bottom sediments until the next rains. Species that live for less than a year are called annual fishes and live in both Africa and South America. Most are in the order Cyprinodontiformes, a group that includes killifishes, topminnows, and rivulines (see "How long do female fishes hold eggs in their body?" in chapter 6).

Many desert fishes, and others that live in low oxygen environments, breathe using more than their gills to maximize their uptake of oxygen. They absorb oxygen via dense blood vessels near the surface of the mouth (electric eels, swamp eels), digestive tract (lungfishes, bichirs), throat (snake-

Pupfishes occur in some of most inhospitable environments possible, areas subjected to drought and extreme temperatures. The Devils Hole Pupfish (*Cyprinodon diabolis*) lives in one small spring in Nevada and is protected under the Endangered Species Act. Photo from U.S. Fish and Wildlife Service

heads), gill chambers (clariid catfishes, gouramis), or gas bladder (featherfin knifefishes; see "How long can a fish live out of water?" and "Can fishes breathe air?" in chapter 2).

Some fishes that live in habitats that dry up seasonally pass the dry season in a type of resting state. They bury in the bottom as their habitat evaporates. Walking catfishes (Clariidae) bury as deep as 3 meters (10 feet) in sandy sediments as water levels drop. African lungfishes (Protopteridae) burrow into mud, secrete a cocoon around themselves, and become inactive in the dry mud until the next rains, an event for which they can wait 4 years. The Australian Salamanderfish (*Lepidogalaxias salamandroides*, Galaxiidae*)* burrows into bottom sediments and surrounds itself with a thick mucous coat. Burying is helped by a bendable neck, a rarity among fishes. Salamanderfish conserve water by absorbing it from the surrounding soil until soil moisture approaches zero.

Because desert fishes live in low numbers and in extreme habitats that push their capabilities in the best of times, these fishes are very sensitive to disturbances such as introduced species, pollution, or water withdrawal. Many desert fishes have been exterminated or require protection and are included in lists of endangered species.

Do fishes live in caves?

About 136 species in 19 families of bony fishes—including five of the six North American cavefishes (Amblyopsidae) and some characins, loaches, minnows, and catfishes—occur in caves around the world, almost entirely in fresh water. Regardless of locale, cavefishes have evolved similar adaptations to the darkness and low food availability that characterize underground habitats.

Typical cavefishes lack scales, color, and eyes. To make up for their loss of eyes, which would be useless in the dark, they have evolved improved hearing and chemical senses such as taste and smell. As a result, they take

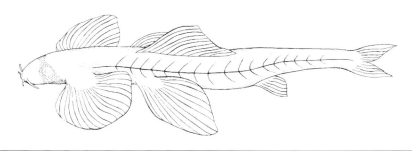

Living in caves has selected for a number of weird-appearing adaptations among fishes. This 3 centimeter (1.2 inch) cave-dwelling Torrentfish (*Cryptotora thamicola*, Balitoridae) from Thailand occupies fast-flowing water. It is eyeless, colorless, and scaleless, as are many cave dwellers. But it also shows adaptations for swift water, including large pectoral and pelvic fins with suction capabilities and a steeply sloping forehead. See the BBC Planet Earth video on "Caves" for live footage. Drawing by S. Madsden in Helfman et al. (2009); used with permission of Wiley-Blackwell

on a weird appearance: eyeless, colorless, scaleless, with large heads and expanded lateral lines.

The biology of fishes that live in caves is tied strongly to the rigors of cave life. These fishes typically move and grow slowly, mature slowly, have few young, and occur in small numbers. Cavefishes are very sensitive to pollutants and introduced predators and competitors, and many are endangered because of these impacts.

How do fishes survive the winter?

In large North American lakes, fishes tend to move to deeper water as temperatures drop in the fall. Centrarchid sunfishes and Northern Pike (*Esox lucius*, Esocidae) occupy deeper water and swim farther offshore under ice. Some species, such as Common Carp and Bigmouth Buffalo (*Cyprinus carpio*, Cyprinidae; *Ictiobus cyprinellus*, Catostomidae) form groups in traditional areas in relatively deep (5 to 7 meters; 16–23 feet) water. Smaller minnows remain in shallow, nearshore areas and occupy piles of twigs, small cracks in rocks and logs, or even bury themselves 0.5 meters (1.5 feet) down in gravel bottoms. As any ice fisher knows, many fishes feed actively once thick ice covers a lake; smelt, numerous salmonids, esocids (Northern Pike, Chain Pickerel), percids (Yellow Perch, Walleye, Sauger), and centrarchid sunfishes can all be caught icefishing. One species that occurs in fresh water at high latitudes across the Northern Hemisphere, the Burbot (*Lota lota*), spawns in winter under the ice.

In ponds and small lakes, however, conditions grow quite harsh during the winter. As ice and snow build up, the water contains less and less oxygen. Fishes that can survive these conditions, such as mudminnows (Umbridae) and pike and Yellow Perch (Percidae), live just under the ice where it is thinnest and where oxygen can diffuse through from the air above. These fishes swim with their noses in contact with the ice, or find gas bubbles and inhale water from around the bubbles. Mudminnows will even swallow air bubbles that have been exhaled by aquatic mammals such as beavers and muskrats.

In streams, salmon, trout, and minnows switch from feeding by day to feeding at night. During the day, they hide between and under boulders and cobbles where the current is weakest.

Many temperate marine fishes leave shallow water kelpbeds when the kelp dies back in winter. Common names of many fishes suggest seasonal movement. Summer Flounder (*Paralichthys dentatus*, Paralichthyidae) spend warmer months nearshore along the coastline and in bays but migrate offshore in the fall to deeper water. In contrast, Winter Flounder (*Pseudopleuronectes americanus*, Pleuronectidae) migrate to deeper water in the summer and then return to bays as the water cools.

Fishes that live in permanently cold oceans, such as polar regions, have special adaptations for dealing with freezing conditions. Many have antifreeze compounds in their blood that keep it from freezing. Because cold water contains more oxygen than warm water, these fishes have very different blood cells. These fishes are sometimes called "white blooded" or "bloodless" because their blood lacks hemoglobin and myoglobin, the compounds that make blood and muscles red. Many of these fishes live nowhere else, such as the Antarctic thornfishes, cod icefishes, and channichthyid crocodile icefishes, plunderfishes, and dragonfishes (suborder Notothenioidei).

Do fishes get sick?

Fishes suffer from many diseases caused by parasites, viruses, fungi, protozoans, and bacteria. They also suffer from decreased organ function due to pollution. The list of parasites that attack fishes is long. Internal parasites include many types of worms. Most are harmless but some such as nematode worms infest Hawaiian Whitespotted Toby (*Canthigaster*, Tetraodontidae) and fill up the pufferfish's body cavity. The infestation causes the body to swell greatly, uses much of the fish's energy, and causes a delay in sexual maturation.

External parasites can be embedded worms, attached leeches, or sea lice (a type of copepod). A few sea lice on an adult salmon probably do little harm. But when several attach to migrating juvenile salmon, as happens when juveniles swim near net pens holding infected farmed salmon, a few lice can be fatal. Population crashes of Pink Salmon (*Oncorhynchus gorbuscha*) in British Columbia are thought to have been caused by sea lice picked

Juvenile Pink Salmon (*Oncho-rhynchus gorbuscha*) with three parasitic copepods ("sea lice") attached. Such infections occur when young salmon migrate past salmon farms and pick up parasites from caged fish. The paired white structures sticking up from one of the sea lice are egg cases. Photo courtesy of Alexandra Morton, Raincoast Research Society

up as young salmon left their birth rivers and swam past aquaculture pens full of infected Atlantic salmon.

Trout in many countries suffer from whirling disease, caused by a single-celled organism that attacks the brain. Whirling disease causes the fish to swim in tight circles and eventually stop moving. This parasite has caused the death of 90% of the fish in some popular trout fishing streams.

A one-celled plant-like organism called *Pfiesteria* (feestaireya) multiplies in bays that receive organic pollutants such as waste water from hog and chicken farms. *Pfiesteria* attacks the skin of fishes, eating the flesh and increasing a fish's risk of other diseases. *Pfiesteria* is one kind of algae that causes "harmful algal blooms" that lead to major fish kills around the world.

The list of diseases of aquarium fishes is also long. Common ailments include ich, tumors, dropsy, fin rot, popeye, hole-in-the-head disease, swim bladder disease, and so on. Ich, also called white spot disease, is caused by a single-celled organism that attaches to the gills and skin of a fish, multiplies, and eventually kills the fish. Fortunately, many aquarium diseases can be treated by a variety of readily available medicines that are applied to the water, preferably after an infected fish is isolated from its tankmates.

Fishes that grow up in polluted waters often have crooked bodies, frayed or missing fins, and damaged gills. Acid rain caused by pollutants from industry and automobiles kills young fishes. Acid rain has eliminated trout from streams and lakes in Sweden and the Appalachian Mountains of the United States. Young minnows raised in water containing large amounts of silt, as occurs when soil erodes from deforested land, have gills that develop poorly and are clogged with mucus. These minnows grow more slowly than minnows raised in cleaner water.

How can you tell if a fish is sick?

Fishes have many ways of telling us that something is wrong. Some parasites such as ich are obvious, causing white spots all over the body of the fish. Other signs of illness include loss of color, swollen bellies, difficulty staying upright, unusual swimming patterns, gulping at the surface, scraping quickly against rocks, rapid gill movement, mucus covering the gills or body, loss of fins, reddened skin, and lying motionless on the bottom.

Are fishes good for the environment?

Fishes play important roles everywhere they occur, as predators, herbivores, prey, transformers and transporters of nutrients and other material, engineers, and when living cooperatively with other species.

- Predatory fishes live in just about every habitat, feeding on invertebrates, other vertebrates including fishes, and an occasional mammal. Predators keep prey populations in check, preventing the prey from decimating their own food resources. Predators on sea urchins on tropical and temperate reefs, mostly wrasses (Labridae), prevent urchin populations from exploding and eliminating algae important as food for many fishes and invertebrates. Predators also perform an important evolutionary function by consuming diseased, deformed, slow, and otherwise poorly adapted individuals.

- Herbivorous fishes—minnows, characins, catfishes, cichlids, damselfishes, parrotfishes, surgeonfishes, rabbitfishes—eat algae, rooted plants, leaves, and seeds, especially in the tropics. On coral reefs, where herbivorous fishes such as parrotfishes and surgeonfishes have been eliminated by overfishing, algae covers the reef, smothering and killing the coral that is habitat for many other animals. In the Amazon River basin, seed-eating relatives of piranha (*Brycon*, *Colossoma*; Characidae) chew off the outer husk of tree seeds or swallow and digest seeds, carrying them up and downstream, thus helping seeds spread and germinate.

- Fishes form the prey base of many food chains in both marine and fresh water. Herrings, anchovies, Capelin, smelt, Eulachon, silversides, minnows, small mackerels, and sandlances are just a few of the superabundant "baitfishes" eaten in vast quantities by sea birds, whales, seals and sea lions, sharks, salmons, tunas, billfishes, and other predators. When prey fishes decline in number due to overfishing or environmental change such as shifting ocean currents, populations of the many seabirds and other animals dependent on them also crash.

- Fishes transform and transport energy and nutrients between different parts of a habitat. Blacksmith (*Chromis punctipinnis*, Pomacentridae) in California kelpbeds feed on floating animals (zooplankton) by day and rest in bottom crevices at night. While resting, they digest their food and defecate on the bottom. Their feces are eaten by gobies, clinids, shrimps, hermit crabs, amphipods, snails, and brittlestars, which are then eaten by larger fishes. Nitrogen excreted by the blacksmith is taken up by the kelp, helping it grow and provide habitat for the many kelpbed animals. Similarly, on coral reefs, juvenile grunts rest over coral by day and move to grassbed areas at night to feed on invertebrates. When they return to the reef, they digest their food and excrete nutrients that promote the growth of algae that live inside corals and are essential for coral growth. Corals with grunt schools resting over them grow faster than corals that lack grunts.

Even in death, fishes play a crucial role in some ecosystems. Salmon migrate from their birth streams to the north Pacific, feed, and grow. When they return to the stream to spawn, they are eaten by eagles, bears, mink, and otters. After they die, bears drag them into the forest and eat them or leave them to decompose. The rotting salmon and bear feces stimulate plant growth inside and along the streams. Large trees along Alaskan streams with salmon grow faster than trees living near streams without salmon. Salmon thus transport nutrients thousands of kilometers (miles)—from the oceanic Pacific Ocean to hundreds of miles up rivers and streams, influencing multiple ecosystems along the way.

Fishes do not only move energy and nutrients around in aquatic ecosystems. They may also act as "ecosystem engineers," contributing to the geology of an area. Parrotfishes eat coral and grind it up in their pharyngeal teeth. As they move about the reef, they excrete the sand, changing its distribution. Other fishes crush clams and sea urchins (e.g., stingrays, emperors, wrasses, surgeonfishes, triggerfishes, puffers), turning the shells into sand particles. Sand tilefish (*Malacanthus*, Malacanthidae) move coral chunks and shell fragments and pile them around their burrows. Tilefish mounds may be the only piles of hard substrate in large areas of sand. Many small fishes (juvenile damselfishes, drums, butterflyfishes, angelfishes, and surgeonfishes) take up residence inside these piles. Male minnows in North American streams (*Nocomis, Semotilus, Exoglossum*; Cyprinidae) construct nests from stones. A chub nest can contain thousands of stones as large as one centimeter (0.4 inch) across, carried several meters from other parts of the stream. The nest again creates hard habitat in areas that are made up of shifting sand and mud. Other minnows besides the chub spawn in the nest, and some will not spawn unless a male chub is present. After the nest is abandoned, aquatic insects inhabit it.

Many fishes live cooperatively with other species. Such "symbiosis" (living together) is seen everywhere fishes occur. Anemone fishes and anemones, shrimps and gobies, grunts and corals, sharksuckers and sharks, and cleanerfishes and their hosts are a few examples. In each example, both members of the pair benefit from the interaction in terms of accelerated growth, increased feeding opportunities, being cleaned of parasites, and keeping predators away. Some of these partnerships are mandatory for the fish: anemonefish are never found without anemones, and shrimp-goby pairs always occur together, never apart. In this way, fishes create healthy environments because the presence of one species makes life possible for the other (see chapter 4, "Are fishes social?").

Chapter 6

Reproduction and Development of Fishes

How do fishes reproduce?

Most fishes reproduce sexually, males fertilizing the eggs of females. Most fish individuals are one gender throughout life, either male or female (minnows, catfishes, salmons, black basses, perchlike fishes, tunas). Although it is often hard (for us) to tell the genders apart, in many species the difference is obvious, especially during the breeding season. Regardless, reproduction in fishes takes many twists and turns along paths not followed by amphibians, reptiles, birds, or mammals. Some fishes reverse sex, from male to female or from female to male. Individuals of other species are both male and female, including some that can self-fertilize.

Sex reversal occurs in at least 34 fish families. It is most common in marine fishes, especially on coral reefs. The prevailing pattern is for an individual to mature first as a female and then later switch to male (seabasses, wrasses, parrotfishes, gobies). In most species with this pattern, males fight over females, and large, dominant males fertilize the eggs of many females. Small males are by comparison unsuccessful reproducers, but every female mates. Hence there is little cost to being small and female, and little benefit to being small and male, which helps explain why the switch occurs. Cleaner wrasses (*Labroides*, Labridae) form harems of one large male and a peck order of up to 10 females. The largest female mates most with the male, on down the list. If the male dies, the largest female changes to male within two weeks.

In species that mature first as male, little fighting occurs over females (moray eels, loaches, snooks, porgies, threadfins, some damselfishes). Because larger females produce more eggs, the bigger the animal, the greater the advantage in being female. Small males still mate successfully and thus incur minimal cost; they wait until they are much larger to switch to female. Male-first species include the popular anemone or clownfishes (*Amphiprion*, Pomacentridae). Anemone fishes live in groups of two large and several small individuals in an anemone. Only the two largest fish are sexually mature, the largest one being female and the next largest being male. If the female dies, the male changes sex to female and the next largest fish in the group quickly matures as a male. If *Finding Nemo* had been true to life, Nemo's dad, Marlin, should have become Nemo's mother shortly after his original mother was eaten by a barracuda. Explain that to your little brother.

Hermaphroditism is the condition of being a functioning male and female at the same time. In hamlets (*Hypoplectrus*, Serranidae), fishes form long-lasting pairs. During spawning, individuals switch back and forth, one fish fertilizing the eggs of the other, then vice versa. One species of topminnow (*Kryptolebias marmoratus*, Rivulidae) is the only fish capable of fertilizing its own eggs. Self-fertilization produces clones of genetically identical offspring.

A few other unusual sexual patterns occur in livebearers and deep-sea anglerfishes. In livebearers such as some mollies (*Poecilia*, Poeciliidae), females mate with males but the male's sperm only stimulate egg development without contributing any genes. All offspring are copies of the mother. Deep-sea anglerfishes (Ceratioidei) do not change sex but females may be 10 to 60 times larger than males. The males are small and parasitic, permanently fused to the side of the female, connected to her bloodstream. The female provides food and oxygen via her blood, and the male provides sperm when the female spawns. Finding a mate in the deep sea is challenging given the vast space involved, so this arrangement assures males and females are close (very close) when the time is right.

Do all fishes lay eggs?

Most fishes (in fact, 98% of bony fishes) lay eggs. In about 2% of bony fishes and half of sharklike fishes, fertilization is internal in the female's reproductive tract and young are born rather than hatched. Males deposit sperm in females via a penis-like structure that is a modified fin (pelvic fins in sharks, anal fin in guppies and goodeids). Embryos use nutrients from the yolk that was originally part of the unfertilized egg (many sharks, coelacanths, scorpaenid rockfishes). In others, including goodeids, bythitid brotulas, and embiotocid surfperches, the mother provides nutrition for the young in addition to the yolk.

In some sharks (e.g., lamnid White Shark), this nutrition involves a form of cannibalism. After using up their own yolk reserves, developing young eat unfertilized eggs or siblings. Many sharks have a placenta very similar to the

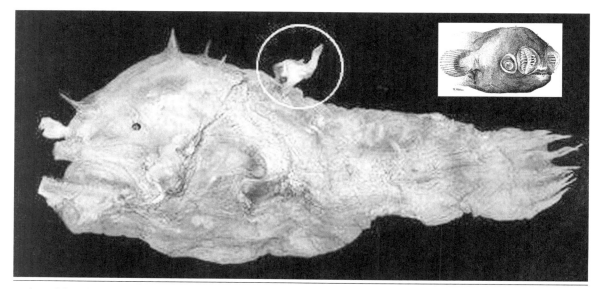

Male and female deep-sea anglerfishes differ in size and biology more than practically any other vertebrate. A 6.2 millimeter (0.25 inch) parasitic male *Photocorynus spiniceps* (Linophrynidae) (circled) is permanently fused to the back of a 46-millimeter (2 inch) female. Males of this species may be the smallest known sexually mature vertebrate. (Inset) a free living, 18 millimeter (0.75 inch) male of the Illuminated Netdevil (*Linophryne arborifera*, Linophrynidae), showing the greatly enlarged eyes and olfactory organ thought to be used in locating females that are four times longer.
Photos courtesy of T. W. Pietsch

one found in mammals, with nutrition coming from the mother. Young are born with a placental scar (yes, many baby sharks have a belly button).

Some live-bearing fishes get around the problem of having too little internal room to produce many developing, relatively large young by staggering the beginning of development of different broods. Females of some poeciliid livebearers and three genera of halfbeaks (*Dermogenys, Nomorhamphus, Hemirhamphodon*; Zenarchopteridae) store sperm and then fertilize eggs in batches. One batch of eggs starts developing but the next batch is not fertilized until a week or so later, and so on. In this way, only one batch at a time is taking up much internal room because the more recently fertilized eggs are relatively small. The result of this process is that females can produce many relatively large babies at intervals of a week or so that would not be possible if all developed at the same time.

Why do some fishes lay so many eggs but other fishes lay only a few?

The number of eggs a female lays or produces depends on short-term and long-term factors. The most immediate influence is body size. Within a species, larger females lay more eggs. In some (salmons, cods, Haddock,

Fishes: The Animal Answer Guide

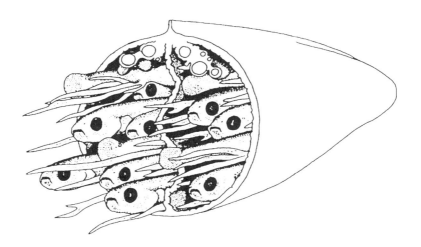

A cutaway view of embryos developing in the Butterfly Splitfin (*Ameca splendens*, Goodeidae). The drawing shows 13 embryos (8 heads and 5 tails) as they sit in the ovary of the mother. The fingerlike projections coming out of the ovary are structures that transport nutrition from the ovary to the developing young. Each embryo is about 3 centimeters (1 inch) long. Illustration by Julian Lombardi, from Wourms, Grove, and Lombardi, 1988; used with permission of the publisher

Striped Bass, flounders), larger females lay larger, better eggs with more yolk, from which larger young hatch.

Ultimately, the number of young a female produces at any one spawning depends largely on their likelihood of survival. The number then is an evolutionary adjustment to past ecological conditions. Species that experience high levels of juvenile mortality tend to produce more and often smaller eggs. Most marine fishes produce larvae that are dispersed widely across the ocean, more than 99% of which die. These species typically produce many eggs. For example, a large female Atlantic Cod (*Gadus morhua*, Gadidae) may spawn 3 million eggs, a large Tarpon 12 million eggs, and a giant Ocean Sunfish (*Mola mola*, Molidae) 300 million!

Fishes in which hatchlings are relatively large or are cared for by the parents until they are large produce few eggs, as in sea catfishes (Ariidae) and Bowfin (*Amia calva*, Amiidae). The amount of yolk in an egg also influences egg number because larger eggs with more yolk take up more room inside the mother. Salmons produce relatively few, yolky eggs, on the order of a few thousand; in some small madtom catfishes (*Noturus*, Ictaluridae), only a few dozen large eggs are laid at one time and then guarded by the male. Coelacanths produce the largest eggs of all, up to 9 centimeters (3.5 inches) in diameter and these eggs hatch in the mother giving birth to live babies. Clutch size (number of young) is between 5 and 26. Live bearing fishes typically produce few young at any one time, again because live young take up more room inside a female than do eggs but they are born at a relatively large size that gives them a greater chance of survival.

Reproduction and Development of Fishes

How long do female fishes hold eggs in their body?

In most fishes, eggs are produced and released in one spawning season, which takes a few months between early egg development and spawning. At higher latitudes and cooler climates, most fishes have only one spawning season per year in both fresh and ocean waters (centrarchid sunfishes, Yellow Perch, pickerels, gobies, kelpfish, Yellowtail). On coral reefs, fishes may spawn repeatedly throughout the year (wrasses, damselfishes), although some species are seasonal spawners (seabasses, snappers, rabbitfishes). Tropical freshwater fishes spawn seasonally, but seasons are defined by rainfall rather than temperature, and fishes spawn as rivers rise during the rainy season.

Live-bearing fishes hold their young for a month on average after fertilization. This time span varies greatly among species. The longest known "gestation" periods are among sharks, with Spiny Dogfish (*Squalus acanthias*, Squalidae) carrying young for up to 2 years, and Basking Sharks (*Cetorhinus*) perhaps as long or even longer. Coelacanths (*Latimeria*) are also thought to have a 3-year gestation period.

Once eggs have been spawned, it is usually only a matter of days before they hatch and swim about freely as larvae. Time until hatching varies greatly, being as little as 1.5 days in Asian carps, 3 to 4 days in centrarchid sunfishes, 1 or 2 weeks in North American and European minnows, a month in sculpins, and 2 or 3 months in Burbot, whitefishes, salmons, and Atlantic Cod. The longest incubation times occur in tropical freshwaters, where cyprinodontiform killifishes and rivulines (e.g., *Nothobranchius*, *Aphyosemion*), many of which are popular aquarium species, live for only 8 months. They spawn, deposit eggs in the bottom, and die. The eggs do not hatch until the next rainy season, which may be only 4 months away but can take as much as 5.5 years.

Where do fishes lay their eggs?

Eggs are scattered on the bottom among rocks (minnows, suckers); dug into sand or gravel (salmons, grunion); placed in a nest made of sand, pebbles, bubbles, or vegetation (minnows, sticklebacks, centrarchid sunfishes, Siamese Fighting Fish); released into the water (cods, wrasses, tunas); or deposited on vegetation or floating objects (herrings, flying fishes, Yellow Perch). Males fertilize the eggs while they are being laid or shortly after.

A few fish species use live invertebrates as a spawning site to protect the eggs against abundant egg predators. Snailfishes (*Careproctus*, Liparidae) lay their eggs inside the gill chambers of crabs. Bitterlings (*Rhodeus*, Cyprinidae) use freshwater mussels, a female depositing eggs into the gill

Callichthyid catfishes, such as this *Corydoras semiaquilus*, are easy to keep in aquariums, serving as useful scavengers that clean up tank bottoms. Their breeding behavior is unusual, females drinking the sperm of males, passing the sperm live through their digestive tract, and then fertilizing their own eggs with the male's sperm. Photo by Stan Shebs

chamber of the mussel, followed by a male who releases sperm over the inflow tube of the mussel.

In some African cichlids, females lay eggs on the bottom but quickly snap them up in their mouth. The female then nips at the anal fin of the male who releases sperm, fertilizing the eggs in the female's mouth, where the eggs develop and are protected. In South American armored catfishes (*Corydoras*, Callichthyidae), a popular aquarium fish, the female places her mouth over the vent of the male and drinks his sperm. She passes the sperm rapidly through her digestive system. She then releases her eggs and holds them between her pelvic fins, and releases the male's sperm to fertilize her eggs. The eggs are then deposited on the river bottom.

Do fishes lay their eggs at the same time and in the same place every year?

Spawning times of many fishes are highly predictable, depending on day length, water temperature, rainfall, and moon phase. Egg laying in seasonal spawners, discussed earlier, can therefore be predicted. Many fishers take advantage of this predictability, expecting the fish to come together in spawning groups as water temperature rises in the spring in temperate lakes or as a particular month approaches on coral reefs.

Predictable spawning locations are best known among coral reef fishes, including seabasses, snappers, croakers, wrasses, parrotfishes, and surgeonfishes. Seabasses and groupers (Serranidae) spawn year after year at the same places, fish traveling more than 100 km (60 miles) to gather on the spawning grounds. In parrotfishes and wrasses (Labridae) and surgeonfishes (Acanthuridae), the exact same coral heads will be used as traditional spawning sites, year after year, by different individuals.

Do fishes breed only one time per year or once in their lives?

Some fishes spawn only once in their lives, all at once. Salmons, lampreys, anguillid (freshwater) eels, and osmeriform southern smelts (Retropinnidae) and galaxiids of Australia and New Zealand are examples. Most fishes, however, spawn several times during a spawning season and several times during their lives (e.g., sharks, lungfishes, sturgeons, gars, tarpons, minnows, trouts, codfishes, seabasses).

What is a baby fish called?

The general name for baby fish is "fry." Fish that have just hatched from eggs and are swimming around are usually called "larvae," and if they still have their yolk sack attached they are called "yolk-sac larvae." Salmons go through a series of development stages, each with a different name, such as alevin, parr, and smolt. Because many larvae are so different from the adults into which they grow, many fish larvae were thought to be entirely different species. Only when fish were raised in laboratories, or after all the different life stages were collected, did people link larvae with their adults, as in the gibberichthyid gibberfishes, whose larvae were once in their own family, the Kasidoridae. The larvae of several groups are still called by their older, family or generic names: *amphioxides* for lancelets, *ammocoetes* for lampreys, *leptocephalus* for all eels, *querimana* for mullets, *rhynchichthys* for squirrelfishes, and *acronurus* for surgeonfishes.

Are all the eggs in the nest full siblings?

The relatedness of fish in a nest very much depends on whether one or both parents care for the young. In fishes that form pairs and where both mother and father take care of the eggs and fry, the young are full siblings with the same parents (cichlids, catfishes, wolf eels). Partial exceptions are seahorses and pipefishes (Syngnathidae). In these fishes, it is the male who becomes "pregnant" and cares for the young. Females lay eggs on the male's belly or inside a pouch on the male's belly. The male then cares for the eggs until they hatch out as baby seahorses or pipefishes. These fishes tend to be very faithful, mating with just one other individual for life.

When only one parent guards the nest, relatedness tends to vary. In fishes, it is usually the male that guards the eggs. A nest may then contain

Many baby fishes do not look anything like the adults of their species and were thought to be different species or even families. This "kasidoron" larva of a Gibberfish (*Gibberichthys pumilus*, Gibberichthyidae) was once placed in its own family, the Kasidoridae. The larva's body is about 8 millimeters (0.3 inches) long, not counting all the trailing filaments. Inset is an adult, about 9 centimeters (3.5 inches) long. Photo courtesy of G. D. Johnson

eggs from several females (darters, Percidae; sticklebacks, Gasterosteidae). In cardinalfishes (Apogonidae), the male broods the eggs in his mouth, but the eggs may come from several females. In some fishes, females prefer to lay their eggs in nests that already have eggs (Fathead Minnow, Threespine Stickleback, Painted Greenling, River Bullhead, Tessellated Darter, Browncheek Blenny). Male chubs (*Nocomis*) build pebble nests where female chubs, as well as females from other minnow species, deposit eggs, and the male guards all the eggs. Thus for these species—along with others that "dump" their eggs in the nest of another species (gars and minnows in sunfish nests, Golden Shiner in Bowfin and Largemouth Bass nests, cichlids in bagrid catfish nests)—the male guards young that are not full siblings but even belong to different species.

The attractiveness to females of nests that already contain eggs leads to surprising behavior on the part of males. In sticklebacks, a male will steal eggs from the nests of other males and deposit the eggs in his own nest, making it more attractive to new females.

Other surprises occur. Even some live-bearing fishes give birth to young fathered by several males. A Bluntnose Sixgill Shark (*Hexanchus griseus*, Hexanchidae) that washed ashore near Seattle, Washington, contained 80 embryos. DNA tests showed that eight different males had fathered the babies. DNA testing also reveals that nests of fishes that usually contain full siblings have a few eggs with different fathers. These "aliens" exist because other males will rush into a nest when a pair is spawning and quickly deposit sperm, thus fertilizing a few of the eggs (centrarchid sunfishes, salmons, pupfishes, wrasses).

Reproduction and Development of Fishes

Once fish leave the care of their parents, they are often dispersed widely by currents before settling into their final habitats. It was generally believed that once this happened, siblings became separated. However, some recent studies using new genetic techniques have found that siblings may stay together much longer. Unicornfishes (*Naso*, Acanthuridae), European Eel (*Anguilla anguilla*, Anguillidae) and Kelp Bass (*Paralabrax clathratus*, Serranidae) are among the species that have been shown to stay together. Future research will undoubtedly discover more.

How is the sex of a fish determined?

In fishes, the sex (or gender) of an individual is determined by its genes, its environment, or both. In most fishes, determination is entirely genetic, each individual becoming male or female solely depending on its genes. In many fish species, the genes determining sex are on sex chromosomes, similar to the XY chromosomes in mammals (skates, anguillid and conger eels, tetras, many catfishes, salmons, killifishes, livebearers, sticklebacks, cichlids, gobies). In mammals, including humans, the male has both X and Y chromosomes, whereas in fishes either the male or female is XY, depending on species.

Environmental factors that can influence sex determination include temperature, age or size, and social interactions. In Atlantic Silverside, (*Menidia menidia*, Atherinopsidae), larvae spawned in spring at low temperatures usually become female; those spawned in summer at higher temperatures become male. Males also result from higher development temperatures in some minnows, gobies, silversides, loaches, rockfishes, cichlids, and flounders. But sex determination, like so many reproductive characters in fishes, is hardly constant. Higher temperatures result in females in (some) lampreys, salmons, livebearers, sticklebacks, and seabasses.

The seabass family (Serranidae) is huge (maybe 475 species), and even within this group, very different patterns occur. Many seabasses mature first as female and then later change to male when they reach a certain size. In these species (such as Nassau Grouper and Red Hind), all of the largest individuals will be male. Fishing that targets the largest fish can then cause a population crash, because too few males are around to fertilize the eggs of females.

In other species, social conditions determine sex. Individuals change from one sex to the other depending on the balance of males and females in the population. In sex-changing wrasses (Labridae), the biggest fish are male. Smaller fish are female and stop growing until a large male dies, then the biggest female changes to male. An opposite situation occurs in anemonefish, where the biggest fish in an anemone is female, the next largest a

In seahorses and pipefishes, the male carries the developing young in a special brood pouch on his belly, and females compete for access to males. This very pregnant male is about to give birth. Courtesy pics.64hd.com

male, and all the others are immature. Bump off the female and everyone moves up a notch, with the largest male changing to female.

Do fishes care for their young?

Many fishes are good parents. About one-third of fish families include species that care for their young. Parenting ranges from guarding a nest containing eggs or young to brooding eggs and young in the mouth to producing body substances that nourish the young.

It is the male that usually does the work. Males construct and maintain nests, chase potential predators away, fan the eggs with their fins to keep oxygen concentrations high and get rid of silt, remove dead or diseased eggs, accompany young that are foraging and protect them from predators, and produce mucus that prevents growth of bacteria on eggs or that the young eat.

Some males go to extremes to care for their young. In seahorses and pipefishes (Syngnathidae), females lay their eggs on or in a brood pouch on the male's belly, where the young develop. The pregnant male helps the developing young with salt excretion, oxygen uptake, and some nutrition until they reach a relatively advanced stage of development. In nurseryfishes (Kurtidae), males develop a downward-bent hook on their foreheads where the eggs are attached and carried until hatching. The Spraying Characin or Splash Tetra (*Copella arnoldi*, Lebiasinidae) deposits its eggs out of water

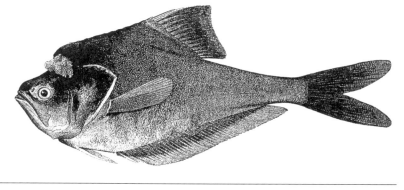

A 15 centimeter (6 inch) male Australian Nurseryfish (*Kurtus gulliveri*, Kurtidae) from Northern Australian streams carries eggs on a hook that projects from its forehead. How the eggs get there remains a mystery.

Drawing from Weber, 1913.

on the undersides of leaves, as much as 10 centimeters (4 inches) above the water surface. The male guards the eggs and keeps them moist by splashing water on them with flips of his tail.

In cichlids and catfishes, the mother or father or both participate in parental duties. Many cichlids and catfishes hold eggs and young in their mouths. Mothers flick their fins to attract free-swimming fry when danger approaches, at which time the mother inhales the young into the safety of her mouth. The Cuckoo Catfish (*Synodontis multipunctata*; Mochokidae) of Lake Tanganyika, Africa, takes advantage of this mouth-brooding behavior. It lays its eggs on the bottom as a female cichlid is picking up her own eggs in her mouth. The young catfish are protected by the female, grow faster than the baby cichlids, and eventually eat some of the cichlid fry that are their "nestmates."

Some African cichlids get help with babysitting. These "non-parental care givers" are usually young from a previous breeding period. They remain with the parents and feed and defend new young or defend and maintain the territory. In turn, they may someday inherit a good nesting spot (*Lamprologus, Neolamprologus, Julidochromis*).

How fast do fishes grow?

The usual pattern of growth in fishes is for very rapid growth early in life, followed by slower—but continual—growth later. This pattern of continued growth differs from other vertebrates (amphibians, reptiles, birds, mammals) that reach a maximum size and stop growing. Although a minnow will never grow as large as a marlin, both will get larger every year as long as their needs for food, clean water, oxygen, and other necessities are met. If conditions are particularly good, growth will be faster.

Growth does however slow and even stop occasionally. Growth is greatly slowed or even halted during winter or drought. If a fish is attacked by a predator and injured, or is stressed by pollution or competitors, or caught

Fishes: The Animal Answer Guide

Second summer's growth

Second winter's growth

First summer's growth

First winter's growth

Core

You can tell a fish's age from its scales. This scale is from a Chinook Salmon (*Oncorhynchus tshaw-ytscha*) caught during its second summer of growth. Winter growth bands are slower and closer to-gether, forming darker regions, whereas summer growth lines are further apart where growth is faster. Adjacent winter and summer growth bands taken together equal one year's growth and are referred to as an annulus (from "annual," or one year). Photograph courtesy of Oregon Department of Fish and Wildlife

and released by a fisherman, growth will stop while repairs are made and the individual's physiology returns to normal. During the reproductive season, fishes stop putting energy into body growth and instead invest in eggs, sperm, fighting over territories or mates, defending young and territories, etc. Again, body growth slows or stops until breeding activities are over.

How can you tell the age of a fish?

It is relatively easy to tell how old a fish is because of the continual growth described in the preceding question. As they grow, fishes add layers of bony substances and proteins to their scales and bones. A fish's scales can therefore be read and the fish aged much like a tree can be aged by its rings (except you usually need a microscope to count growth rings on fish scales). Where distinct warm and cold seasons occur, such as in North America and Europe, a fish adds a pair of rings to its scales and vertebrae and ear bones and fin spines every year. Each ring pair appears as dark lines and light lines. The dark lines are where the added bony layers are close together due to slow growth, such as in winter. The light lines are the layers that are added much faster, such as during the summer. Dark lines of slow growth are also added during the breeding season. These are called "spawning checks" and can be used to tell how many times a fish bred.

It is more difficult to determine the age of a fish from the tropics where temperatures are fairly constant because all the lines on the scales look the same. In tropical areas that experience wet and dry seasons, distinct dark

Reproduction and Development of Fishes

(dry season) and light (wet season) lines may be added. But if drought and rainfall occur irregularly, the lines are not very useful. And because growth slows as a fish gets older, the growth lines on the scales get closer and closer together, making counting more difficult. As a result, we can know how old young fish are but our counting becomes less accurate as a fish gets older.

It may surprise you to know that the age of very young fish can be known in days, not just years. This precision is possible because young fish lay down layers to their ear bones every day. These layers can be seen under very strong microscopes such as electron microscopes. Electron microscope techniques also allow us to sample the chemicals in these rings, which can then be used to determine where the fish lived during the earliest part of its life. Different habitats contain different chemicals, leaving behind a chemical history of sorts. The biography of a young fish can almost be read like a book.

Fishes: The Animal Answer Guide

This display of cichlid fishes at the Georgia Aquarium in Atlanta is representative of what one might see diving in Lake Malawi, Africa. As many as 900 cichlid species occur in this lake and nowhere else. Other large lakes in Africa have similarly large flocks of cichlid species, again occurring nowhere else.

African lungfishes such as this *Protopterus* are legendary for their ability to estivate, burrowing into mud and remaining inactive for years while waiting for rains to return. African tribespeople take advantage of this natural means of preservation by digging up lungfish, burrow and all, and saving them for a later date as a source of fresh fish. Photo courtesy of L. and C. Chapman

A dead male Sockeye Salmon (*Oncorhynchus nerka*) decomposing in a stream near Juneau, Alaska. Pacific salmon die after spawning, and their remains provide nutrition for a variety of stream-dwelling as well as terrestrial plants and animals.

The head of this 20 centimeter (8 inch) male Bluehead Chub (*Nocomis leptocephalus*) develops a fleshy hood, blue color, and hard, sharp bumps as it builds a pebble pile nest, defends its territory, and courts females. Photo courtesy of Paul Vecsei

Spawning in many fishes involves a single female surrounded by several males. (A) California Grunion (*Leuresthes tenuis*) spawn on beaches at the top of the tide zone. A single female (shown by arrow) buries her tail in the sand with her head up and lays eggs. Males encircle her and release sperm. (B) Robust Redhorse Suckers (*Moxostoma robustum*) spawn in groups over gravel. In this photo, the female (shown by arrow) has one male on her left and two on her right. Panel A photo courtesy of M. Horn; panel B courtesy of B. Freeman

Parrotfishes get their name from their huge, parrot-like jaws that are used to bite and scrape coral to feed on the algae inside. This is the head of a Rainbow Parrotfish (*Scarus guacamaia*) from the Caribbean, one of the larger species.

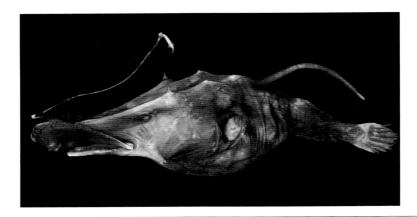

Anglerfishes are a diverse group of 11 families that live in the deep sea. This 15 centimeter (6 inch) long female Wolf-trap Angler (*Lasiognathus amphirhamphus*) possesses a lure at the tip of her modified first dorsal spine, the business end of which is just above the two spines behind her eyes. The rodlike structure pointing back is the other end of the dorsal spine. The rod slides in a groove on her head, allowing the anglerfish to move it forward when fishing but to retract it otherwise. Photo and lure explanation courtesy of T. W. Pietsch

Hagfishes, otherwise known as slime eels or slime hags, get their name because of the huge amounts of mucus they produce: one disturbed hagfish can fill a 2-gallon bucket with slime in a matter of minutes. The slime serves many purposes, such as making them relatively inedible and thus not as desirable as prey, at least to other fishes. Photo by J. L. Meyer

The Humphead or Napoleon Wrasse (*Cheilinus undulatus*) is the largest member of one of the largest families of fishes. Growing to over 2 meters (6.5 feet) and 150 kilograms (330 pounds), Humpheads feed on hard-bodied prey such as cowries, boxfishes, and seastars. These exceedingly desirable wrasses have been overfished on reefs throughout the Indo-Pacific region and are now classified as endangered. Photo by David Hall / seaphotos.com

Fishes grow accustomed to divers and in many places accept handouts. Here a diver off the coast of British Columbia has cut open a sea urchin and is using it to coax a 2.1 meter (7 foot) long male Wolfeel (*Anarrhichthys ocellatus*) out of his den. Photo by David Hall / seaphotos.com

Piranhas aren't that deadly. Although certainly capable of tearing things up, piranhas are primarily scavengers, or at least most of the reported attacks on humans involved drowning victims.

The Asian Arowana or Golden Dragonfish (*Scleropages formosus*). Overcollecting for the aquarium trade pushed this species to the brink of extinction. Desirable color varieties have sold for as much as $5,000. Photo by Marcel Burkhard

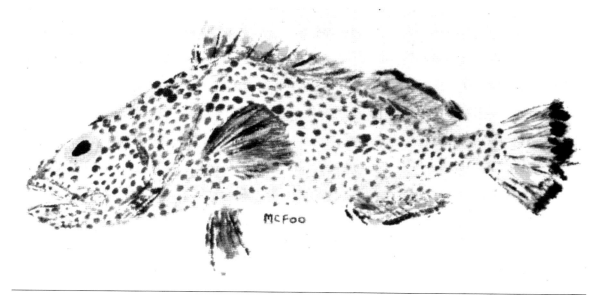

Gyotaku print of a Red Hind (*Epinephelus guttatus*). The fish, about 20 centimeters (8 inches) long, was first covered with black acrylic ink and pressed on a cotton t-shirt. Colors were then added with acrylic paints, matching the real color as much as possible. Artwork by W. N. McFarland

Oscars (*Astronotus ocellatus*) are a popular aquarium fish from the Amazon region of South America and are members of the huge family of cichlid fishes. Oscars are graceful and attractive and have been bred to produce a variety of colors. Unlike most cichlids, Oscars have large mouths, eat smaller fishes, and grow large, even in small aquariums.

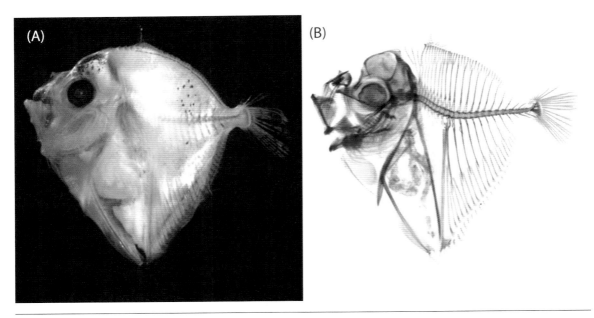

To examine the internal anatomy of small fishes like this 5.8 millimeter (0.25 inch) Moonfish (*Mene maculata*), ichthyologists clear and stain specimens so that the flesh becomes transparent, bones are stained red, and cartilage is stained blue. (A) Preserved specimen; (B) specimen after clearing and staining. Prepared and photographed by G. David Johnson

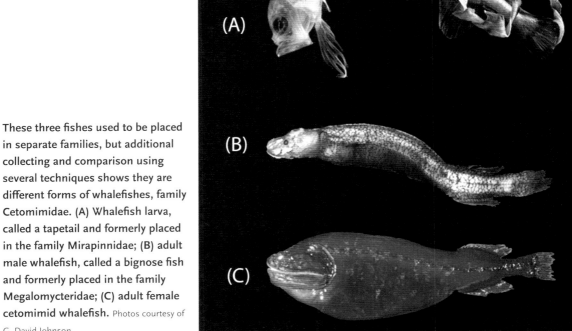

These three fishes used to be placed in separate families, but additional collecting and comparison using several techniques shows they are different forms of whalefishes, family Cetomimidae. (A) Whalefish larva, called a tapetail and formerly placed in the family Mirapinnidae; (B) adult male whalefish, called a bignose fish and formerly placed in the family Megalomycteridae; (C) adult female cetomimid whalefish. Photos courtesy of G. David Johnson

Chapter 7

Fish Foods and Feeding

What do fishes eat?

Taking the 31,000-plus fish species worldwide, you could say that fishes eat everything. Some fish species are in fact omnivorous generalists, eating vegetation, insects, other fishes, detritus, zooplankton, you name it. Such generalist feeding habits help explain the amazing success of a few species that have been introduced in many places. Common Carp (*Cyprinus carpio*, Cyprinidae) and bullhead catfish (*Ameiurus*, Ictaluridae) are two freshwater examples. Many other fishes are opportunistic feeders, preferring certain foods but able to take advantage of what is available. Bluegill and Pumpkinseed Sunfish (*Lepomis*, Centrarchidae) eat zooplankton, aquatic insects, larval fish, snails, and just about any other animal that they can fit in their mouths.

In contrast, many if not most fishes are more particular about what they can and will eat, specializing on certain food types. Their anatomies reflect these specializations. The freshwater cichlids of South America and Africa, numbering perhaps 2,000 species, include species that specialize on sponges, sediments, algae, leaves, mollusks, insects, phytoplankton, zooplankton, fish scales and fins, fish eyes, fish eggs and embryos, and other fishes. Their lips, gill rakers, and jaw and pharyngeal (throat) teeth are especially suited for their preferred food type.

Across the 500-plus families of fishes, the catalog of unusual food types is long, with many surprising entries:

- One feeding type attacks parts of other fishes, such as scales and fins. Fin biters have sharp-edged teeth, whereas the teeth of scale eaters are

Madagascar cichlids are considered to be ancestral to the incredibly diverse cichlids of continental Africa, with their equally diverse feeding habits. Unfortunately, the Pinstripe Damba cichlid (*Paretroplus menarambo*), such as this one. are extinct in the wild, but the species is kept going through captive breeding.

often more file- or rasplike. Fin biters include young piranhas (*Serrasalmus*, Characidae) in the Amazon, saber-tooth blennies (*Aspidontus, Plagiotremus*; Blenniidae) on coral reefs, many African cichlids, some Asian barbs (*Puntius*, Cyprinidae), freshwater pufferfishes (*Tetraodon*, Tetraodontidae), and bumblebee gobies (*Brachygobius*, Gobiidae). Scale eaters include several African cichlids and leatherjackets (*Oligoplites*, Carangidae) on coral reefs. Cookie-cutter sharks (*Isistius*) sneak up on tuna, marlin, and dolphins and take a single, circular, half-dollar size plug out of their victim's sides. Unfortunately, several of these so-called "partial consumers" are also popular aquarium fishes, causing no end of headaches for aquarists.

- A number of fishes both in marine and fresh water pick parasites from the body, mouth, and gills of other fishes. Specialist cleaners include the cleaner wrasses (*Labroides*, Labridae), remoras (Echeneidae), and some butterflyfishes (*Chaetodon*) in the tropical Pacific, and neon gobies (*Gobiosoma*) in the tropical Atlantic (see "Are fishes social?" in chapter 4). Many other fishes pick parasites part-time, otherwise feeding more generally on small invertebrates (juvenile damselfishes, angelfishes, butterflyfishes, surfperches, wrasses, and centrarchid sunfishes).

- Fishes in many habitats feed on the feces of other fishes, an activity called "scatophagy" or "coprophagy." Scats (*Scatophagus*), a popular Asian aquarium fish, get their scientific name from this behavior. Large *Labeo* minnows in Africa accompany hippos, using their excrement as a constantly supplied food source. Coprophagy is common on coral reefs. A study in Palau, Western Caroline Islands, found 45 different species of reef fishes from eight families (mostly sea chubs, damselfishes, wrasses, rabbitfishes, surgeonfishes, and triggerfishes) eating feces from 64 fish species in 11 different families, preferring the feces of planktivores over waste products from herbivores.

Fishes: The Animal Answer Guide

- A few fishes specialize on, or at least supplement their diet with, the eyes of other fishes. A narrow-bodied cichlid in Lake Malawi, called the Malawi Eyebiter (*Dimidiochromis*) does not make a good aquarium fish because of its eye-popping activities. A North American cyprinid, the Cutlip Minnow (*Exoglossum maxillingua*) will also remove eyes from its tankmates.
- Some African cichlids specialize in attacking female cichlids that are carrying young in their mouths (see "Do fishes care for their young?" in chapter 6). These babynappers repeatedly ram the head of the mother until she spits out a few young, which are then snapped up by the predator.

More usual food types include small plants or animals in the water column (phytoplanktivores and zooplanktivores), other fishes (piscivores), plants (herbivores), and detritus (detritivores). Many fishes that normally feed on live prey will not hesitate to scavenge on dead things.

Fishes that swim above the bottom and feed on zooplankton have a mouth that can shoot forward and thereby greatly enlarge. This increase in mouth volume creates strong suction pressures that cause small, swimming prey to be vacuumed into the round tube of the jaws. Such "pipette" mouths occur in both freshwater and marine zooplanktivores, including Bluegill Sunfish, herrings, damselfishes, fusiliers, bogas, wrasses, and cichlids, to name a few.

Piscivores come in three general types. Lie-in-wait predators rest on the bottom and are usually well-camouflaged or even bury themselves with just their eyes exposed (flatfishes, stonefish, scorpionfishes, lizardfishes, weeverfishes). These ambush predators wait for a prey fish to swim by close enough to be captured with an explosive rush. Frogfishes and anglerfishes speed up the process by waving a small lure just above their mouths (see "How do fishes find food?" below). Ambush predators generally inhale their prey with large mouths or impale prey on needle-like teeth, in the case of lizardfish.

Most other piscivores pursue their prey. Pursuers fall into two groups, burst swimmers and active chasers. Burst swimmers come from a wide variety of fish groups but all have evolved a similar body shape. They are elongate and have long, tooth-studded jaws, with their fins placed far back on their bodies. They hover in the water and dash at prey with a single burst of speed, impaling their prey with sharp teeth. These are the classic predatory fishes—barracuda, pike, gar—but an amazing variety of unrelated fishes in both marine and fresh water have converged on this body type and feeding method. Their common names often refer to their resemblance to a pike (Esocidae), including pike minnows (Cyprinidae), pike charac-

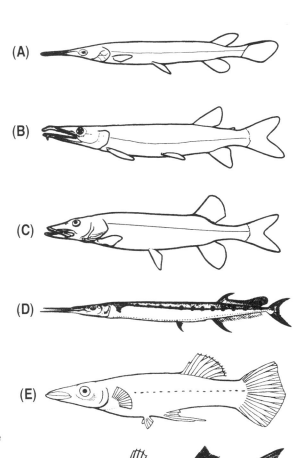

Burst-swimming predators tend toward similar shapes and behaviors, regardless of their ancestry and size. These fishes catch their prey with a single burst of speed, impaling their victims with sharp teeth. They include (A) gars (Lepisosteidae), (B) pike-characins (Ctenoluciidae), (C) pikes and pickerels (Esocidae), (D) needlefishes (Belonidae), (E) Pike Killifish (Poeciliidae), and (F) barracudas (Sphyraenidae). From Helfman et al. (2009); used with permission of Wiley-Blackwell

ins (Ctenoluciidae), pike killifish (Poeciliidae, the guppy family), pikehead (Osphronemidae), Australian Long-finned Pike (*Dinolestes lewini*, Dinolestidae), as well as needlefishes (Belonidae) and some cichlids.

The remaining group of predators actively chases down their prey, pursuing it for more than an initial burst of speed. Body shapes vary but most are fairly robust and streamlined, and capture usually involves overtaking prey by swimming rapidly, inhaling prey with an expandable mouth, or immobilizing prey with a weapon. Many are also popular game fishes. Examples include the Largemouth and Smallmouth Black Bass (*Micropterus salmoides* and *M. dolomieu*, Centrarchidae); snooks (Centropomidae); numerous cichlids that look very much like a black bass or snook (e.g., Peacock Cichlids in South America, *Cichla ocellaris*, also called Peacock Bass) and *Rhamphochromis* cichlids in Africa; many seabasses (Serranidae); tunas (Scombridae); and mahi-mahis (Coryphaenidae), to name just a few. The weapon wielders are the various billfishes and are probably the fastest fishes in the sea (e.g., marlins, Sailfish, spearfishes, Istiophoridae; and Swordfish, Xiphiidae).

Regardless of attack mode, many piscivores rely on an extremely expandable pipette mouth that creates strong suction. Frogfishes (Antennariidae), Stonefish (*Synanceia*), black basses (*Micropterus*), dories (*Zeus*), slingjaw wrasses (*Epibulus*), and cornetfish and trumpetfish (*Fistularia*, *Aulostomus*) are examples. Both suction pressure and forward shooting of the mouth aid in capture. During an attack, the mouth becomes an extension of the body that can be shot out faster than the fish can swim through the water. Mouth volume may be increased by 15 to as much as 40-fold at the critical moment.

A major factor determining what a carnivorous fish can eat is simply food size. Most fishes cannot eat anything they cannot swallow. Swallowing involves first getting something into the mouth and second getting it down the throat. Hence fishes are "gape-limited." Gape limitation is less a challenge for the few exceptional fishes that can actually tear up or chop up their food into swallowable bites (Tigerfish, piranhas, Bluefish, barracudas, and sharks). But most fishes do not have strong enough jaws and large, sharp teeth with cutting edges. If a fish miscalculates prey size and attacks a prey item too large to swallow, it can choke to death. It is not unusual to find a Largemouth Bass with a Bluegill Sunfish sticking halfway out its mouth or a pickerel with a stickleback caught in its throat, floating at the surface of a lake, dead.

As a result, and although larger fishes can eat larger prey because they have larger mouths, many very large fish eat remarkably small items. Yellowfin Tuna (*Thunnus albacares*, Scombridae) weighing 23 kilograms (50 pounds) or more may have their stomachs packed with larval fishes, each measuring a few millimeters (a fraction of an inch) in length. This is why an angler can often catch large gamefish with what seem to be very small lures. The notable exceptions to the "smaller-than-your-mouth" size rule are found in the deep sea, where many fishes have hinged jaws and expandable stomachs and can swallow prey longer than themselves. Feeding opportunities in the deep sea are few and far between, so fishes there must be able to take advantage of anything that comes along.

Many fishes that lack cutting teeth still overcome gape limitation by breaking up prey, but they do it in an unexpected manner. They shake, twist, and spin their food until it comes apart, much as you would twist taffy or separate a chicken leg from the bird's body. The best spin-feeders are "eels," including the freshwater eels (Anguillidae) but also many other eel-like fishes closely or distantly related to the anguillids (moray eels, Muraenidae; snake eels, Ophichthidae; conger eels, Congridae; swamp eels, Synbranchidae; pricklebacks, Stichaeidae; gunnels, Pholidae).

Even large predators are gape-limited. Most fishes can only eat things they can swallow whole. Large groupers, such as this Giant Grouper (*Epinephelus lanceolatus*), that grows as large as 2.7 meters (9 feet) long and weighs up to 600 kilograms (1,320 pounds), must swallow prey whole and hence cannot attack and dismember large prey. It is the exceptional predators with sharp, cutting teeth that are able to chop up their prey.

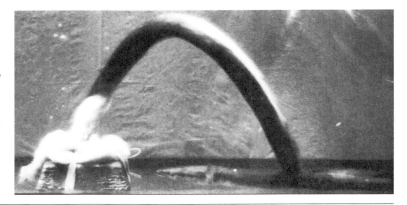

An American Eel (*Anguilla rostrata*) spins while feeding. The 50 centimeter (19 inch) eel has grasped the bait, a fish filet attached to a lead weight, in its mouth and has begun to rotate. Twist in the body is evident: the light underside of the head and the dark back both face the camera at the same time that the light belly is facing upward.

These and other "rotational feeders" have one thing in common, namely a long, thin body that lacks hard fin spines or large paired fins such as pectorals or pelvics. This body type allows them to twist and spin rapidly, up to 14 rotations per second, while holding onto prey (the best human ice skaters cannot rotate faster than about five or six rotations per second). Other long-bodied or eel-like animals also spin to dismember prey, including some reptiles (i.e., crocodiles engaged in their famous death roll, some snakes) and a few amphibians (sirens and caecilians). The cookie cutter sharks mentioned earlier use the same spinning behavior to remove divots of flesh from their much larger, living prey. These small (50 centimeter, 20 inch) predators are also called cigar sharks because of their relatively long bodies and small fins.

Moray eels (Muraenidae) are among the best spinners and add another tactic also made possible by having an eel-like form. While holding onto a prey fish with strong jaws and sharp teeth, they tie a knot in their tail and pass the knot forward until it presses against the prey fish. As the knot presses and the jaw holds, the prey is torn apart. Hagfishes, also elongate, use this knot tying technique to remove chunks of skin and flesh from their already dead meal. Hagfishes (Myxinidae) lack true jaws and teeth and instead hold onto their prey with the sucking disc at the front of their body.

Plant-eating herbivores do not have to fool their prey or chase it down, but they do have to overcome mechanical and chemical defenses. Bony fishes can eat microscopic algae (shads, herrings, Milkfish, Blackfish, mullets, cichlids, minnows), scrape or nibble algae off rocks (Chinese Algae Eater, minnows, catfishes, cichlids, sea chubs, blennies, pupfishes, damselfishes, parrotfishes, surgeonfishes), or eat whole plants or take pieces out of leaves or digest seeds (Asian minnows, cichlids, characins, halfbeaks). However, plants give up their nutrients begrudgingly due to their hard, leathery, or rubbery exteriors; extremely tough cell walls; placement of more nutritious parts inside inedible packages; or noxious chemicals throughout their tissues. Herbivorous fishes rely on vision, taste, and digestive capabilities to separate the readily edible from the tough and sometimes toxic. Herbivores are almost entirely daytime feeders, pointing out the importance of vision in food choice.

Tooth type (see "Do all fishes have teeth?" in chapter 2) and intestinal tract anatomy are adapted to plant type. Leaf eaters generally have broad, flattened jaw teeth suited for nipping off the ends of algal fronds or leaf tips. Herbivorous fishes often have grinding stomachs and very long and convoluted intestinal tracts (animal flesh is relatively easy to digest and predators consequently have relatively short intestines). Herbivores use pharyngeal mills or highly acidic stomachs to break down the cell walls of plants. Unlike insects and many herbivorous mammals, fishes generally lack intestinal bacteria that help digest plant matter via fermentation. The few exceptions include surgeonfishes (Acanthuridae) and sea chubs (Kyphosidae). Sea chubs feed heavily on brown algae that are avoided by most other herbivores.

An extreme among herbivores that points out the challenges presented by eating algae involves the parrotfishes (Labridae), which feed on the algae that grow on and often inside corals. To get at the algae, parrotfishes have massive beaklike teeth and equally massive pharyngeal jaws. Their jaw teeth scrape or bite off chunks of coral rock. Coral fragments are then passed back to the pharyngeal jaws where the coral is ground up. The fish is not after the coral animals but the algae that grow inside the coral skeleton. The pharyngeal grinding breaks up the cell walls of the algae. In the process, coral is ground down into sand. As parrotfishes swim about the reef, they excrete clouds of sand along with feces. On some reefs, abundant parrotfish are significant producers and movers of sand.

The ability to eat a wide variety of foods was a major step in fish evolution. The living members of the more primitive bony fish groups (lungfishes, coelacanths, sturgeons, paddlefishes, bichirs, gars, Bowfin) are all carnivores, as are all sharks and the jawless hagfishes and lampreys. Most are primarily fish eaters, habits that undoubtedly reflect the feeding behavior of their ancestors (lungfishes eat mollusks and North American paddlefish feed on zooplankton). It is only in the more recently evolved teleosts (modern bony fishes) that make up 99% of living fish groups that we see a diversified diet that includes plant life in its various forms. This ability greatly expanded the habitats in which fishes could live as well as moving fishes down to the most productive levels of the food chain.

Do fishes chew their food?

Unlike reptiles, fishes can chew their food. Unlike mammals, most fishes chew not with their jaw teeth but instead use throat teeth. The teeth in these pharyngeal jaws are well adapted to handling the kind of food a fish normally eats. Mollusk feeders have large, flattened or rounded pharyngeal teeth best suited for crushing hard-bodied prey. Fish eaters have long, curved pharyngeal teeth that pierce the prey and also prevent it from escaping back out the mouth. Insect feeders have smaller, pointed pharyngeal teeth.

The mouths of most fishes are therefore primarily the "capture" tool of fishes, chewing and other processing of prey occurring in the pharyngeal jaws or even in a gizzard-like structure in the stomach. Gill rakers (bony, fingerlike projections on the inside of each gill arch) also help to keep prey from escaping out the gill openings while food is being moved to the stomach. Plankton feeders have fine, closely spaced gill rakers that help in the capture of tiny animals, whereas piscivores often have stout, prickly rakers that keep larger prey from escaping.

Although most of the teeth on the various bones are pretty much the same, a few exceptional fishes have different shaped teeth in different parts of the mouth. For example, wolffishes (Anarhichadidae) feed on hard-bodied prey such as sea urchins. Wolffishes are more like mammals than like other fishes in having biting and grasping teeth in the front of their jaws and crushing teeth toward the back, analogous to our canines and molars.

Some bony fishes replace their teeth as they grow. Mammals have two sets of teeth, baby or milk teeth when young, and adult or permanent teeth

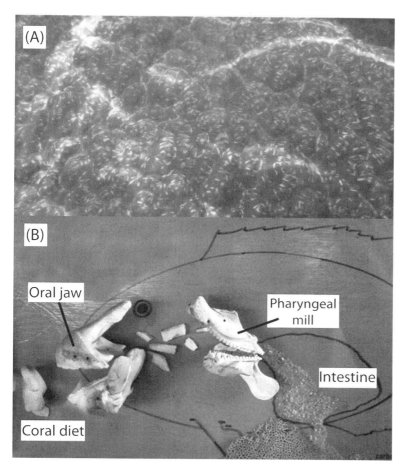

(A)

(B)

Oral jaw

Pharyngeal mill

Intestine

Coral diet

Parrotfishes bite coral and scrape the surface of coral heads, then grind coral bits into sand in the process of digesting the algae inside. (A) Bite marks on a coral head from parrotfish grazing. (B) A model showing how coral fragments are ground into sand as they pass through the mouth, pharyngeal mill, and gut of a parrotfish.

later. Many fishes, by contrast, replace their teeth throughout the lives. Functional teeth sit on the jawbones while replacement teeth lie under them, embedded deep in the jaws. Replacement teeth erupt when the outer teeth are lost or shed. Sharks are best known for the multiple rows of replacement dentition, but many bony fishes also have teeth-in-waiting (e.g., Bowfin, characins such as piranhas, salmons and trouts, surgeonfishes, gobies).

Some large predators such as seabasses, Largemouth Bass, snook, and marlin that feed on active swimming prey actually have relatively small teeth. Their teeth feel like sandpaper and would seem incapable of capturing much. Instead, these fishes rely on a large mouth that when opened rapidly creates strong suction pressures. Hence many fishes capture their prey by vacuuming them in, close their mouth once the prey are inside, use their various mouth teeth to manipulate prey so they can be swallowed head first, and then pass the prey back to their pharyngeal jaws where the real action takes place (billfishes skip much of this process by whacking or spearing prey with their lethal swords and swallowing the helpless victim).

How do fishes find food?

Fishes use all their senses, including one we do not have, to find food. Vision plays an important role in food finding, but many fishes forage at night or live in permanent darkness (the deep sea, caves, very murky water) and must rely on other means of locating food.

Even fishes that initially locate food via nonvisual means rely on vision in the final attack phase of feeding. Most zooplanktivores locate and track their prey visually, even though their prey is very small. Daytime feeders can see and follow extremely small objects because of the high density of visual receptor cells (retinal cones) in their eyes. Having many small cones also makes their eyes exceedingly sensitive to movement. Their eyes therefore have high acuity (point-to-point resolving power) and high motion detection. Predators such as tarpon, trout, black bass, barracuda, pompano, and tuna rely on motion detection and bright flashes of light reflecting off the silvery sides of many bait fishes to tell them where prey are. This behavior explains why fishing lures that are cast and retrieved or trolled often have silvery, reflective surfaces.

Surprisingly, many night-feeding fishes rely on vision, especially fishes that swim in open water and feed on zooplankton. Their visual tasks are made somewhat easier because the invertebrates and fish larvae they feed on are larger than the plankton animals that swim about by day. Fish groups includes sweepers (Pempheridae), squirrelfishes (Holocentridae), cardinalfishes (Apogonidae), and bigeyes (Priacanthidae) on coral reefs; some surfperches (Embiotocidae) in kelpbeds; and Black Crappie (*Pomoxis nigromaculatus*, Centrarchidae) in temperate lakes.

Nocturnal feeders have very large eyes. Many also have a reflective layer at the back of their eyeballs, the "tapetum lucidum" (from the Latin, meaning "bright tapestry")," which causes light to pass through the eye twice, thereby increasing light sensitivity. A tapetum is the same structure that causes eye shine in cats, dogs, crocodiles, raccoons, and deer. In addition, the receptor cells (retinal rods) in their eyes are very large and thus able to capture any available light. Such light sensitivity comes at a cost because large rods sacrifice acuity and motion detection for sensitivity. Several predators feed at night and also have very large eyes (e.g., most sharks and chimaeras; coelacanths; sturgeons; Walleye and Sauger, Percidae).

Special situations require special adaptations. A few fishes spend a good deal of their foraging time out of water, such as mudskippers (*Periopthalmus*, Gobiidae), which move across exposed mudflats and climb up plant roots to find food. They have a strongly curved cornea and slightly flattened lens that allows them to focus on objects in air. The Four-eyed Fish (*Anableps*, Anablepidae) swims with half of its eye out of the water, searching for insects. Its eyes are split in half horizontally, each having two pupils and a retina that is divided into top and bottom sections. Cavefishes and deep-sea fishes that live in perpetual darkness have very small eyes or have lost their eyes completely over the course of evolution. Plants do not grow in the dark and prey animals are scarce in both caves and the deep sea. Fishes in these habitats save energy by not having eyes, but they instead use sound and smell to find food.

Water is an excellent transmitter of sound; sound travels 4.5 times faster through water than through air (1,500 meters /5,000 feet per second in water versus 330 meters/1,100 feet per second in air). As a result, sound travels great distances and is a good source of information. Fishes are very sensitive to sounds, especially at the low frequencies that travel farthest. Few of the sounds important to fishes occur at higher frequencies. Most fishes cannot hear sounds above 500 Hz or cycles per second, whereas human hearing goes up to 20,000 Hz. The exception is some herring relatives (Clupeidae) that are able to hear sonar clicks at 100,000–180,000 Hz (100,000–180,000 cycles per second) produced by the dolphins that feed on them.

Although fishes cannot hear frequencies as high as humans can, fishes can tell which direction a sound came from, a useful ability for finding food and avoiding predators. Anyone who has snorkeled or scuba dived knows that humans are unable to tell the direction a sound comes from. Try it the next time you are in a pool.

Fishes hear with both their lateral line and their inner ear. The lateral line is a series of nerve cells along the side and head of a fish that are sensitive to water movement. Sound waves traveling through water, such as vibrations caused by a swimming prey fish, move the water and stimulate the lateral line receptor cells or the "otoliths" (ear bones) of the inner ear. Many fishes (minnows, suckers, characins, catfishes, herrings, cods, squirrelfishes) use their gas bladder as a sounding board that vibrates and amplifies sounds, sending a sensory signal to the inner ear and increasing sensitivity to sound. Minnows and their relatives (otophysan fishes) possess a series of small bones derived from modified vertebrae called "Weberian Ossicles" that connect the anterior end of the gas bladder to the inner ear. These bones pivot end-to-end, turning a small vibration at the gas bladder end into a greater vibration at the head end. These fishes have the highest sensitivity and greatest frequency range of hearing among non-clupeid fishes.

Fishes use hearing to find prey, turning toward the source of a sound and moving in that direction until they find the sound producer. A swimming fish actually leaves a sonic trail of disturbed water behind it. Lake Trout (*Salvelinus namaycush*, Salmonidae) can detect and follow prey fishes

Four-eyed Fish (*Anableps*) frequently swim at the water's surface with the "terrestrial" half of their eyes poking out of the water as they search for insects. Wikimedia Commons, http:// en.wikipedia.org/wiki/Anableps_anableps

in total darkness by listening to the water disturbance noises left behind by their swimming prey. Several predators, especially sharks, are attracted to the squeaks, squeals, grunts, thumps, and buzzes produced by injured fishes. Many fishing lures attract gamefish because they make a sound like swimming or injured fish.

Smell and taste are related senses that rely on detection of chemicals. Fishes are sensitive to a wide variety of chemicals but respond most strongly to proteins and amino acids. Most fishes have nostrils to detect odors. The sense of smell (olfaction) is used in finding mates, recognizing predators, directing migrations, and finding food. Taste buds are used more often for foraging. Taste buds occur on the tongue, as in mammals, but many fishes, especially catfishes, have taste buds all over their body, fins, and especially on their whisker-like barbels. Goatfishes (Mullidae) have muscular barbels with which they probe the sand in search of invertebrates. Sea robins (Triglidae) have taste buds at the tips of their pectoral fins.

Fishes have a sixth sense that humans lack. Fishes are sensitive to bioelectric fields. Many fishes can find prey by simply detecting the electric field the prey creates due to muscular contractions, nerve impulses, or just because the salt concentration of a fish's body fluids differs from the salt concentration of the water around it. Sharks, lungfishes, coelacanths, paddlefishes, sturgeons, eels, salmon, trout, catfishes, tunas, and many other fishes are electrosensitive. Electrosensory organs are scattered across the skin of these fishes and are often concentrated in the head region. Paddlefish (*Polyodon spathula*, Polyodontidae), which feed on zooplankton, use their electric sense to detect planktonic animals, an ability that was discovered just recently. The elongate and flattened snout of this fish contains

Fishes: The Animal Answer Guide

Goatfishes occur in coral reef areas around the world. They are frequently seen over sandy areas probing for invertebrates with their chin barbels (muscular whisker-like appendages) and then push into the sand to snatch up a prey item. Inset photo (upper right) shows the prominent barbels of the Caribbean Yellow Goatfish (*Mulloidichthys martinicus*, Mullidae) foraging in the larger photo. Main photo by Doug Bronski; inset photo courtesy of Cecil Berry

many electric receptors. It acts as an antenna that allows Paddlefish to detect individual zooplankters 9 centimeters (almost 4 inches) away.

Sharks and rays are so sensitive to electric fields that they could detect the 1.5 volt output of a D-cell flashlight battery several miles away if it were not for electrical interference from the earth's geomagnetic field. Sharks find prey buried in the sand by homing in on the bioelectric output of the prey. Sharks also dive into the sand after buried, active electrodes. Sharks can sense the electrical output from a swimming human 2 meters (6 feet) away.

Some amphibians (and duckbill platypuses) are also sensitive to electric fields. However, one thing that fishes do that no other group of organisms can do is create an electric field and then detect and identify objects that enter into the field. Special electricity-generating cells made from modified muscle cells produce the field. Electricity-sensing cells are made from modified lateral line cells. Three families of tropical freshwater fishes possess this complete electrolocalization sense: gymnotid knife fishes of South America, elephantfishes (Mormyridae) of Africa, and upside-down catfishes of Africa (Malapteruridae, Mochokidae; see "What are electric fishes?" in chapter 2).

Many fishes simplify the task of finding food by bringing it to them instead. Luring is most highly evolved in the anglerfishes and their relatives (goosefishes, frogfishes, batfishes; order Lophiiformes). Many of these bizarre-looking animals live in the deep sea, where food is scarce and energy conservation is crucial. It is far preferable to lure prey to you rather than having to find and chase it.

Anglerfishes attract prey by dangling a lure just above and in front of their mouth. The lure, called an "esca," is the tip of their first dorsal spine, which is modified to look remarkably like a small fish, shrimp, or worm.

Fish Foods and Feeding

The North American Paddlefish (*Polyodon spathula*) lives in large rivers and filters zooplankton with the aid of its paddle-like snout. The paddle contains numerous electricity receptors that help it find zooplankton swarms.

The lure is wriggled in a lifelike manner, and may even produce a chemical attractant. The predator is well camouflaged to resemble the bottom or the dark surrounding waters. Small fishes approach the lure and are quickly inhaled by a large mouth. Escape is often prevented by long, backward-facing teeth.

Luring is such a successful tactic that it has evolved in many fish groups other than anglerfishes, but especially in other deep-sea fishes. The lure may involve the dorsal spine (as in a scorpionfish, Scorpaenidae) or other body parts. Deep-sea hatchetfishes (Sternoptychidae), lanternfishes (Myctophidae), and stargazers (Uranoscopidae) have lures inside their mouths; barbeled plunderfishes (Artedidraconidae) wave chin barbels (whiskers); chacid catfishes use barbels attached to their upper jaw; snake eels (Ophichthidae) use their tongue as a lure; and in gulper eels (Eurypharyngidae) the tip of the tail glows to attract prey.

An unusual variation on luring and ambush predation involves a predatory cichlid from Lake Malawi, Africa. The Kalingo (*Nimbochromis livingstonii*) is a moderate size cichlid (25 centimeters, 10 inches) that takes advantage of the habit of many small cichlids to scavenge on recently dead fishes. The predator lies on its side on the bottom and assumes a blotchy coloration typical of a dead fish. When scavengers come to investigate and even pick at its body, the predator erupts from the bottom and inhales them. This is the only fish that is known to play possum, except the possum is more like a panther.

Are any fishes scavengers?

Many fishes besides some cichlids will scavenge. Some feed primarily on dead things, but just about any fish—including many herbivores and pred-

Fishes: The Animal Answer Guide

ators—will take advantage of the easy meal presented by a dead or dying animal lying on the bottom. If this were not the case, bait fishing would not be so successful.

Species that make a living primarily by scavenging are sharks, catfishes, freshwater eels (*Anguilla*), and especially the hagfishes (order Myxiniformes). Hagfishes are primitive boneless, jawless, eel-like marine fishes that aggregate in large numbers on anything lying dead on the seafloor. They use their rasping "teeth" to tear away at flesh and organs, sometimes holding on with their mouth and running an overhand knot from tail to head, levering against the body of the food item to tear chunks away. A few hundred hagfish working over a large dead fish can turn it into skin and bones in a matter of hours. The efficiency of scavenging hagfish might cause one to think twice about being buried at sea.

How do fishes eat hard-shelled animals?

Many fishes eat hard-bodied prey such as mollusks (snails, clams) and sea urchins. These "durophages" (as in the phrase "durable goods") usually have very strong jaws with rounded, crushing teeth, often backed up with similar teeth on their pharyngeal jaws. Freshwater Drum (*Aplodinotus grunniens*, Sciaenidae), large suckers (*Moxostoma*, Catostomidae), Shellcracker Sunfish (*Lepomis*, Centrarchidae), and even small stream fishes such as the federally protected Snail Darter (*Percina tanasi*, Percidae) are part of this group. Marine species include many stingrays, Sheepshead (*Archosargus probatocephalus*, Sparidae), and wolffishes (Anarhichadidae) in temperate oceans. On coral reefs, mollusk crushers include wrasses such as the giant Humphead Wrasse (*Cheilinus undulatus*, Labridae) that eats cowries as well as crown of thorns starfish. Not surprisingly, several African cichlids are molluskivores and possess molar-like pharyngeal teeth necessary to crush shells.

Do fishes store their food?

Apparently not. Water is a poor medium for food storage because it carries odors so well. It is difficult to hide a food item from other scavengers, which means just about all other fishes.

Do fishes use tools to obtain food?

Not very often. Tool use involves manipulating something in the environment to achieve a purpose. Although many mammals and birds use tools such as sticks and rocks to probe for, catch, or crush prey, tool-us-

Archerfishes (Toxotidae) are well-known for their ability to shoot down objects above the water, both insects on branches and pieces of hamburger held above the surface by someone . Experiments have shown that one archerfish can watch another and learn how best to aim at a target. Photo by R. Wampers

ing fishes are unusual. The best-known examples come from the incredibly diverse tropical family of wrasses (Labridae: *Halichoeres, Cheilinus, Coris, Thalassoma*). Many wrasses feed on hard-bodied invertebrates such as clams, scallops, snails, and urchins. These fishes have strong pharyngeal jaws, discussed earlier, to crush their prey after it is swallowed. However, some make the task easier by first holding the mollusk in their mouth and bashing it against a rock as if the rock were an anvil.

A few fishes use water jets to acquire food, which is a form of tool use. Archerfish (*Toxotes*, Toxotidae) are the best studied. Archerfish shoot droplets of water by pressing their tongue against a groove in the roof of their mouth. The water bullets can travel 150 centimeters (4.5 feet) and knock insects off branches. Archerfish not only allow for the bending of light rays at the water's surface (refraction) but can also track and shoot down moving prey. In an underwater variation on this tactic, triggerfishes (Balistidae) blow water out their mouths repeatedly at the base of a sea urchin, tumbling the urchin over and exposing its undefended lower half. The triggerfish then tears the urchin apart with its powerful jaws and incisor-like teeth.

Fishes: The Animal Answer Guide

Chapter 8

Fishes and Humans

Do fishes make good pets?

Fishes make excellent pets, and most ichthyologists kept aquarium fishes when they were younger and many still do. In fact, aquarium keeping is one of the most popular hobbies in the world, with $15 to $30 billion worth of fishes and aquarium equipment sold annually. If the right species are purchased from a reputable dealer, fishes are relatively cheap and relatively easy to keep alive for years. A small tank (40–95 liters, 10–25 gallons), an electric pump and filter, lights, and heaters are the major expenses, followed by the fishes themselves, plants (live or plastic), and fish food. Many colorful, active, interesting, and long-lived freshwater species can be purchased for less than a dollar, including tetras, loaches, catfishes, minnows, carps, barbs, killifishes, livebearers, cichlids, labyrinth fishes, and gouramis (see "Which species of fishes are best known?" in chapter 12). These fishes are often raised in ponds (in Florida, California, southeast Asia, or Africa), rather than taken from the wild.

Keeping fish is also a healthy activity for people. In addition to teaching responsibility (but much less responsibility than a cat, dog, or bird), people gain psychological benefits from fish keeping. Watching fish in an aquarium is a relaxing activity, which is why many doctor's and dentist's offices have fish tanks in their waiting rooms.

As with any activity that involves maintaining live animals in captivity, there are drawbacks to fish keeping. While keeping freshwater fishes is relatively easy, marine fishes are another story altogether. Marine fishes are much more sensitive to water quality, temperature, food type, and disease than most freshwater fishes. Additionally, very few marine species can be

A captive bred, marine tropical fish, the Banggai Cardinalfish (*Pterapogon kauderni*, Apogonidae) is among the few coral reef species that can be bred in aquariums. Native to a small region in Indonesia, the species was thought to have been driven into extinction due to overcollecting. But captive breeding and escapes of captive-bred individuals have helped bring the species back. Photo by Jon Hanson

bred in captivity and thus must be captured in the wild, often using methods that destroy habitats and other animals (e.g., use of poisons to capture reef fishes).

These complications drive costs up and also mean that many marine fishes do not live long in a home aquarium. Many problems can be avoided by purchasing fishes that are on the American Marinelife Dealers Association's Ecolist of species that are easily maintained (www.amdareef.com). Fishes should be purchased from dealers certified by the Marine Aquarium Council (www.aquariumcouncil.org).

A popular alternative to buying exotic fish bred or captured from foreign lands is to keep fishes that are native to where you live. In the United States, the North American Native Fishes Association (www.nanfa.org) promotes the sustainable capture, care, breeding, and conservation of fishes native to the United States, Canada, and Mexico. A major concern of NANFA, and of biologists everywhere, is what people do with fishes they no longer want, such as fish that have grown too large for a tank. It is difficult to kill a pet, even a fish, so many people release unwanted fish into rivers, lakes, or the ocean. This careless practice has led to the establishment of introduced predators on and competitors with native fishes, as well as the spread of diseases to native fishes. Tilapia, walking catfish, piranhas, oscars, pacus, snake eels, pleco catfish, and lionfish are just a few examples of released fishes that have become established nuisances. Flushing an unwanted fish

Fishes: The Animal Answer Guide

Loricariid catfishes from South America in the genus *Pterygoplichthys* are popular aquarium fishes, commonly referred to as "plecos" or "plecostomus." Unfortunately, they quickly outgrow their tanks, reaching sizes of 50 to 70 centimeters (20–27 inches). Fish keepers then release the fish into native habitats around the world, rather than killing them. As a result, these fish have become a major nuisance, as in some southern Florida waterways where they construct nest burrows and cause erosion of canal banks. Photo by Pawel Ciesla Staszek

down the toilet is one solution; placing a tropical fish in the freezer is another, more humane practice. Sometimes pet stores will take back fish that they have sold previously, but most stores are unwilling to do so because of concerns about spreading disease to the other fish they are trying to sell.

What is the best way to take care of a pet fish?

The most important thing is knowing the basic biology of the fishes you want to keep. What temperatures are they happiest at? What foods do they like? What fishes do they not get along with? Some fishes require certain kinds of water chemistry (soft versus hard), or places to hide, or plants to eat. The Internet is full of websites written by knowledgeable people and devoted specifically to the kinds of fishes that most people are likely to keep in an aquarium. Many books are available to advise a beginner or an expert on how best to take care of and breed different species. These books are available in public libraries and sold in pet stores and online.

At its simplest, the biggest mistakes people make is allowing water to go bad and feeding too often (the two are often related). The basic rule is to periodically replace one third of an aquarium with water of similar chemistry and temperature to what is already in the tank. Feeding only what the fish can consume in 5 minutes is a good practice; many fish can get along fine without eating every day. Next on the list of common mistakes are overcrowding the tank (again leading to water quality problems and oxygen stress), and keeping fishes together that just do not get along (e.g., that fight with or eat one another).

Fishes and Humans

Do fishes feel pain?

We cannot ask a fish directly if it is feeling pain the way we can ask a person. As a result, whether or not fishes feel pain is a topic debated among biologists (and others). The answer greatly depends on what definition of pain you use, and what kind of behavior you require of the fish. If you want to know if fish feel pain the same way that we do, the answer is probably "no." Our sense of pain involves a part of our brain (the neocortex) that is lacking in fishes.

On the other hand, if you use a definition of pain that can be applied to non-human animals, you get a different answer. Pain is defined (by the International Association for the Study of Pain) as "an unpleasant sensory or emotional experience associated with actual or potential tissue damage." Specifically, for an animal to experience pain, it must (1) have the right body parts (nerves, physiology), (2) behave in way that is not a simple reflex reaction (such as pulling your hand away after touching a hot stove), and (3) be able to learn from the experience so as to prevent it happening in the future.

Several experiments have shown that fish meet all three requirements. Rainbow Trout (*Oncorhynchus mykiss*, Salmonidae) have nerve endings around their mouth and on their head (where we hook them when fishing) that respond to acid and bee venom. These "pain receptors" are six times more sensitive to such substances than skin receptors in people. Many fishes also have nerves and nerve pathways similar to those that function during painful reactions in mammals, including humans. These pathways travel to similar parts of the brain (the hypothalamus) and release the same types of chemicals that are released when we experience pain.

Behaviorally, there is again good evidence that fishes feel pain. Toadfish (*Opsanus*, Batrachoididae) were given electrical shocks. They responded by grunting, which is something they do when disturbed. After a few shocks, they began to grunt just when they saw the shock equipment, before any shocks were given. They had learned to associate the electrodes with the unpleasant experience. When the lips of Rainbow Trout were injected with acid, the trout rubbed their lips on the bottom of the tank, rested on the bottom and rocked back and forth, increased their breathing rate, and stopped feeding. Because these activities did not occur immediately, they cannot be considered simple reflexes. Goldfish (*Carassius auratus*, Cyprinidae) reacted to painful stimuli (unpleasant temperatures applied directly to their skin) with reflexes. But later they avoided the equipment that caused the reaction, so they had learned from the experience and behaved in a way to prevent it happening again.

You be the judge.

Fishes: The Animal Answer Guide

The white triangles, circles, and other shapes show the location of nerve endings on a Rainbow Trout's (*Oncorhynchus mykiss*) head, mouth, and lips that are sensitive to strong touch, harsh chemicals, or high temperatures. Some of these pain receptors are as sensitive as those on the human eyeball. From Sneddon, Braithwaite, and Gentle, 2003

What should I do if I find an injured fish or a fish that looks diseased?

Unfortunately, there is not much we can do about sick fish in the wild. Even a sick or injured fish will swim away if you try to pick it up. Its predators will probably be more successful at capturing it, which is natural selection operating as it should. If the fish is too weak to escape, it will probably die soon, regardless. Bringing it home and trying to nurse it back to health (assuming you can figure out what is wrong, which usually requires a veterinarian) only means releasing it back into the wild. If the fish is diseased, it may spread its illness to other fish.

In the situation where someone sees many sick, dying, or dead fish, it is a good idea to call a local health, police, or natural resource department. A fish kill may be a crime scene because someone is illegally dumping pollutants into the water. Sick fish may also represent a human health hazard, which means the health department will test the fish and perhaps post signs in the area advising people not to eat the fish.

How can I see fishes in the wild?

It is fun to watch fishes from a pier or a dock or the side of a stream. But it is much more fun to watch fishes under water. All you need is a mask and snorkel and maybe swim fins (and a wetsuit in colder water). It is a good idea to first get comfortable breathing through a snorkel in a swimming pool or even a bathtub. Most people catch on quickly and the two-thirds of our planet that is covered with water is theirs to explore.

Snorkeling is cheap and easy but it does limit you to surface waters and is affected by how long you can hold your breath. You are also still looking down on fish most of the time, and most fish are colored in a way that makes them hard to see from above (see "Is there a reason for the color patterns of fishes?" in chapter 3). The next step is to learn to scuba dive. Scuba diving requires taking a course taught by a certified instructor, and lessons and equipment are expensive. But the activity allows you to spend hours up close and personal with the fishes. We heartily recommend scuba lessons for anyone who can afford them.

Regardless of how you enter the fish's world, remember we are land animals. Safety concerns should always be foremost, beginning with being able to swim. Water temperature, currents, waves crashing on rocks, boat traffic, and potentially harmful critters (jellyfish, sea urchins, sharp rocks, stinging corals) are just a few things to worry about. And of course, never swim or dive alone.

If you live in a place where you cannot watch fish in the wild, the next best thing is to visit a public aquarium. And in fact, we do both whenever we can.

Should people feed fishes?

It depends on whether the fishes are in the wild or in captivity, such as in a pond. Although fishes do best on a natural diet, it probably will not do much harm to feed human food to wild fish. Fish scraps, hard-boiled eggs, and bread (unfortunately white bread) are common food types tossed to fishes. The food should be something that is undoubtedly safe to eat, so if you would not eat it, do not feed it to the fish (Avoid prepared foods with lots of artificial ingredients, colors, and preservatives. Those are good for neither you nor the fish).

In many places, people feed fish to attract them to an area so the fish are easier to watch. This is common in underwater parks and reserves, and makes the fish more tame and approachable. Although this is generally a harmless practice, it can cause problems because the fish learn to associate people with a free meal and can become aggressive if you do not bring food. People have been bitten by predators (moray eels, pompano, snappers, barracuda) expecting a handout. Feeding sharks has been outlawed in some places for just this reason.

In some zoos, public aquariums, and fish hatcheries, food dispensers are set up so visitors can buy prepared fish pellets for a small price. This and only this kind of food should be tossed into the pond, and only into ponds where the dispensers are located. Many fish in captivity are on very special diets, fed at special times. The wrong food and overfeeding create prob-

Public aquariums are a great place to see live fishes. These visitors to the Georgia Aquarium in Atlanta are admiring a realistic coral reef exhibit, which is designed to mimic a reef from the Solomon Islands, right down to the live corals, sharks, and many fishes. Many public aquariums promote fish conservation, breed fishes, and educate the public to appreciate the beauty of the natural world.

lems just as they do in a home aquarium. Some displays of very expensive fishes, such as Koi (ornamental Common Carp) ponds, have signs requesting that you throw nothing into the ponds because of their special diet and sensitivity to metals, such as copper pennies. These fishes are there to be enjoyed for their colors and gracefulness. That should be enough.

Chapter 9

Fish Problems (from a human viewpoint)

Are some fishes pests?

A pest is something that causes problems for people, usually because of unnaturally large numbers (ecologists also include plants and animals that are a problem for other plants and animals). Most pest fishes are found in areas that have been disturbed, because numbers seldom get to the pest level in natural communities where predators and disease usually keep populations in check.

Some fishes could be considered annoyances if not pests, even at low numbers. Many of us have had the unpleasant experience of wading in a pond and having sunfishes such as Bluegill Sunfish (*Lepomis macrochirus*, Centrarchidae) picking at hairs, scabs, or moles on our body. This unsettling behavior is really just an extension of the natural feeding activities of the Bluegill. Bluegill are generalist feeders, and one of their roles in lakes is to serve as cleanerfish, removing parasites from the skin of other fishes. The Bluegill is really doing you a favor, not realizing that your leg hairs are not an attached parasitic copepod.

Bluegill and other sunfishes may reach pest levels as a result of a process called "stunting." A stunted population consists of many small fish. It results largely from a lack of predators, usually due to overfishing of sport fishes. When most of the large predators, such as basses, pikes, Walleye, and sturgeons have been removed from a lake, the sunfishes continue to reproduce. Their numbers grow but because of competition for food their food supplies become reduced. As a result, the fish do not grow to large size but still keep on breeding, leading to a lake chock full of small Bluegill or Redbreast Sunfish. Bullhead catfishes and Yellow Perch are other North American native fishes that produce stunted populations.

More commonly, pest fishes are introduced species that do not serve the purpose for which they were introduced. Again, stunting is a common complaint. Goldfish, Common Carp, and tilapia are notorious for stunting. These fishes are introduced to provide food for people but due to stunting wind up being too small to be desirable. Other introduced pest species cause problems because they are predators on native fishes or invertebrates. Walking catfishes, lionfishes, Asian Swamp Eels, Chinese carps, Guppies, Mosquitofish, Ruffe, Round Goby, and snakeheads reach pest status in many places in the United States. Two sci-fi/horror movies were produced in 2004, capitalizing on the discovery that snakeheads had been introduced and were breeding in a Maryland pond. *Frankenfish* portrayed genetically engineered snakeheads. *Snakehead Terror* had snakeheads transforming from pests to predators when human growth hormones were dumped into a local lake.

In other countries, as well as in the United States, introduced sport fishes (Brown and Rainbow trout, Smallmouth and Largemouth bass, Northern Pike, sunfishes, Peacock Bass) are predators that wipe out their prey base and also feed on native sport fishes. The lesson from this list is that introducing any fish that does not occur naturally in an area should only be done after careful study, if at all.

Can there be too many fishes in a lake or river?

Stunted populations, discussed above, are an example of too many fish in one place. Most anglers complain about too few rather than too many fish, except where anglers compete with predatory fishes for sport fishes. For example, anglers in Florida complain about too many gars (*Lepisosteus*, Lepisosteidae) and Bowfin (*Amia calva*, Amiidae) eating the Largemouth Bass that the anglers would rather catch.

From a nuisance standpoint, the physiology of some fishes results in unpleasant conditions for humans. Alewives (*Alosa pseudoharengus*, Clupeidae) entered the Great Lakes through man-made canals in the 1940s. They fed on zooplankton and replaced the native whitefishes and their numbers exploded. Alewives are naturally saltwater fish but can survive in fresh water. However, their physiology is better adapted to salt water. This makes them physiologically stressed and therefore sensitive to sudden environmental changes, such as rapidly reduced water temperature. When Alewives move inshore to spawn in the spring, they are often exposed to reduced temperatures from inflowing streams and cold water upwellings. The result is that

Brown Trout (*Salmo trutta*, Salmonidae) are native to Europe and western Asia but have been introduced throughout the world. They provide good fishing but have wiped out native fishes in North America, Australia, New Zealand, and South Africa, where large predatory fishes did not occur naturally.

they die by the millions and wash up on beaches, where they rot. In Lake Michigan, beaches have to be closed and bulldozed clear of stinking fish. Again, problem fish are more likely to be fish that are not native to an area.

Do fishes kill ducks in ponds and other bodies of water?

Most adult ducks are safe from fish predation, but ducklings are undoubtedly eaten by a variety of large predatory fish such as Largemouth Bass and pike. Muskellunge and large Northern Pike are perfectly capable of eating all sizes of ducks. Goosefish (*Lophius americanus*, Lophiidae), admittedly a marine species, get their name from their presumed goose-eating habits, if that counts. But given that many ducks including mergansers, sawbills, scaup, mottled ducks, goldeneyes, and long-tailed ducks eat fishes, a little turnabout is fair play.

Are fishes dangerous to people or pets?

Very few fish attack people or pets (we will ignore sharks for the time being). Some injuries occur when people accidentally step on or place their hands on spiny bottom fishes that are well camouflaged and do not move (scorpionfishes, stonefishes, weeverfishes, stingrays). But most injuries occur when people handle fishes carelessly, and the fish tries to defend itself with its spines or teeth. Gars, Bowfin, barracudas, Bluefish, halibut, piranhas, tigerfish, and pufferfishes are among fish that will bite and should be handled with care and respect.

Almost all fishes have sharp spines, some that contain venom that produces a painful sting. People are commonly hurt while handling catfishes, pinfishes, rabbitfishes, and even sunfishes. A few fishes can produce a strong electric shock that can stun and even knock a person out (Electric Eel, torpedo rays, electric catfishes), but again the shock is almost always a defensive tactic on the part of a frightened or handled fish.

There's a right way and a wrong way to pick up a stingray, as a student on an ichthyology field trip who missed the demonstration on how to pick up a stingray found out the hard way. Treatment traditionally employs compression to stop the bleeding, heat to destroy the toxin, then ice to reduce the swelling.

Contrary to what scary movies might make you think, piranhas (*Serrasalmus*, Characidae) seldom if ever attack live humans (see "Do fishes bite?" in chapter 4). Most reports turn out to have involved drowning victims, after they drowned. The exceptions involve a recent upsurge in attacks in new reservoirs along the Amazon River. But the fish are not feeding, just protecting their young. Australian tandan or eel-tailed catfish (*Tandanus*, Plotosidae) also have a reputation for attacking waders and swimmers who come near their nests.

A good subject for a factual scary movie would be the behavior of a small pencil catfish, the Candiru (*Vandellia cirrhosa*, Trichomycteridae) of the Amazon River basin. Pencil, or parasitic, catfishes normally eat mucus and scales from other fishes or pierce the skin or gill cavities of other fishes and feed on blood. But the Candiru occasionally makes a mistake that can be an excruciating experience for a swimmer and fatal for the catfish. Candiru are attracted to the small currents that come out of fish gills. They swim "upstream," lodge themselves in the gills with spines that project backward from their gill covers, and feed on gill tissue. A human bather, usually a male, who pees while in the water can attract a Candiru. Documented cases (with photographs) exist of a Candiru stuck in someone's penis, requiring surgery to be removed. Young boys in the Amazon are taught not to stand up and pee over the side of a canoe for fear that one of these fishes will swim up the stream. Being nibbled on by a Bluegill pales in comparison.

Fish Problems

Small but scary, pencil catfish, or Candiru (*Vandellia cirrhosa*, Trichomycteridae) are less than 15 centimeters (6 inches) long, but there are verified accounts of them swimming up a man's penis and lodging themselves there, requiring surgical removal. Photo by P. Henderson, PISCES Conservation Ltd.

Do fishes have diseases and are they contagious?

Few fish diseases can be passed on to humans. However, fishes are intermediate hosts for some parasites that can be transmitted to people and cause severe abdominal pain, vomiting, and diarrhea. These parasites include intestinal tapeworms, liver flukes, and parasitic nematodes, all from freshwater fishes. Almost all can be killed by proper cooking.

A number of harmful bacteria live on fish skin and spines, so when people get stabbed while cleaning fish, the wounds can become infected. At least 12 bacterial species found in fishes can cause infections, including *Salmonella*, *Vibrio*, *Mycobacterium*, and *Streptococcus*. People stabbed while cleaning St. Peter's Fish (*Sarotherodon galilaeus*, Cichlidae) have become infected by *Vibrio* and had to have limbs amputated.

Is it safe to eat fish?

Sad to say, many fish are unsafe to eat. As discussed above, fish that are caught in contaminated water and eaten without thorough cooking can pass disease-causing protozoa, bacteria, and viruses to people. Even worse are fish contaminated with compounds that are not destroyed by cooking. For too many years, we have treated lakes, rivers, and the ocean as convenient places to dump our wastes. As a result, many fish have incorporated these toxins into their bodies. These fish contain toxins that are poisonous to them and us.

Because of our polluting habits, nearly 40% of U.S. lakes and rivers have fish consumption advisories, meaning it is not safe to eat the fish that live there. Fishes in these waters have unsafe levels of mercury, pesticides (DDT, chlordane, diazinon, atrazine, malathion, mirex, toxaphene), herbicides (atrazine), dioxins, polychlorinated biphenyls (PCBs), heavy metals (aluminum, arsenic, cadmium, chromium, lead, selenium), endocrine dis-

A fish consumption advisory sign in coastal Delaware. Fishes here are contaminated primarily with PCBs.

rupters, creosote, and many, many others. Before eating freshwater fish, it is a good idea to check with a local health department to see if a fish consumption advisory has been issued for the place where you are going fishing.

Marine fishes are often not as heavily contaminated as freshwater fishes, although many species that live in coastal lagoons and other nearshore environments are not safe to eat. Again, checking with state and local health departments is a good idea.

Not all problems from eating fish are caused by human pollution. A toxin found in tropical reef fishes around the world is ciguatera. People who eat ciguatoxic fish suffer stomach cramps, heart rate changes, reversal of sensations (i.e., ice cream feeling hot), and possibly death from respiratory failure. The most common culprits are large predators such as moray eels, groupers, snappers, and Great Barracuda, and the toxin is not destroyed by freezing or cooking.

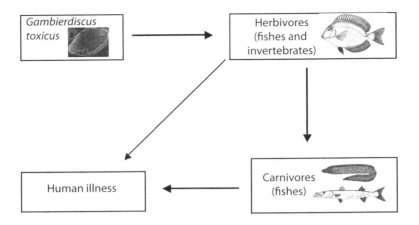

Ciguatera fish poisoning on coral reefs. A one-celled alga called *Gambierdiscus toxicus* produces a potent toxin that passes up the food chain, becoming more concentrated at each step. When people eat contaminated fish, they receive a heavy dose of the toxin. Modified from Helfman (2007); algae photo courtesy of the Florida Marine Research Institute

Ciguatera is a good example of an accumulated, magnified toxin, one that is passed along the food chain and thus concentrated in top predators. The toxin is first produced in certain kinds of reef algae, which are eaten by herbivores, which are eaten by small carnivores, which are eaten by larger predators. People are at the top of this food web and receive a huge dose of toxin by eating a single, large fish. The algae that produce the toxin grow best on recently disturbed surfaces, such as what happens during dredging, dynamiting, and ship anchoring.

Ironically, because the toxin does not break down from freezing, persons far from coral reefs have suffered ciguatera poisoning. Twenty people at a dinner in Calgary, Canada, suffered ciguatera poisoning from eating thawed-and-cooked reef fishes imported from Fiji.

What should I do if I get injured by a fish?

Bites and stab wounds from spines should be treated as any deep cuts by stopping bleeding, cleansing, antibiotics to prevent infection, and so on. Some highly venomous stings, such as those caused by stonefishes, lionfishes, eel-tailed catfishes, and stingrays, can be excruciatingly painful. Professional medical help may be needed because some people are particularly sensitive and can go into anaphylactic shock and stop breathing. For stingray and stonefish stings, first aid manuals suggest applying hot compresses (but not so hot as to burn) to the wound to destroy the proteins in the venom. Call 911 or a poison control center.

Fishes: The Animal Answer Guide

Chapter 10

Human Problems (from a fish's viewpoint)

Give a man a fish and he'll eat for a day. Teach a man to fish and he'll deplete the oceans.
                                                    The Book of Bob, Ironies 24:7

Are any fishes endangered?

Many fish species have gone extinct recently, and many more are in danger of extinction. The official, global list of threatened and endangered species is kept by the International Union for the Conservation of Nature (IUCN). IUCN publishes a Red List of endangered plants and animals that it updates regularly. The most recent list identified 93 fish species that had gone extinct and 1,170 species worldwide that were threatened with extinction. We know this list is incomplete because the best records are from industrial and developed countries whereas most fish species live in tropical, developing countries where record keeping is not as good. It is likely that between 20 and 35% of the world's 11,000 species of freshwater fishes and perhaps 5% of 17,000 marine species are in serious decline or already extinct.

North America has about 1,200 freshwater fish species, of which about 500 to 550 (40 to 45%) need protection. Of the 900 freshwater fishes that occur in the United States, 124 are protected under the Endangered Species Act. Such legal action seldom occurs until a species is close to extinction, which means that many more than 124 are threatened.

At both global and national levels, most endangered fishes live in fresh water, and all known extinctions involved freshwater fishes. These statistics reflect the degree to which we have altered, degraded, and destroyed freshwater habitats and their inhabitants. And although ocean fishes on the whole are not as threatened as freshwater fishes, many marine fishes have declined greatly in numbers.

The main causes of decline and extinction are habitat alteration (disturbance of river and stream bottoms and shoreline vegetation, dam building, water withdrawal, turning natural rivers and streams into channels and ditches), introduced species, chemical and sediment pollution, hybridization, and overfishing. Often, several causes work together to cause declines. Main causes differ in different locales, with habitat disturbance being the chief culprit in fresh water and overexploitation from fishing (and habitat disruption from destructive fishing practices) causing the most declines among ocean fishes. Behind all these problems are human overpopulation and overconsumption.

Which are the most endangered fishes in the world and in the United States? Experts disagree (of course), but there is an official list of species that cannot be traded between countries because of low and declining numbers. Trade is restricted by the Convention on International Trade in Endangered Species (CITES). The list includes nine fishes: African and Indonesian coelacanths (*Latimeria*), Shortnose and Baltic sturgeons (*Acipenser brevirostris* and *A. sturio*, Acipenseridae), Golden Dragonfish (*Scleropages formosus*, Osteoglossidae), Julien's Golden Carp (*Probarbus jullieni*, Cyprinidae), Cui-ui Sucker (*Chasmistes cujus*, Catostomidae), Mekong Giant Catfish (*Pangasianodon gigas*, Pangasiidae), and Totoaba (*Totoaba macdonaldi*, Sciaenidae). The coelacanths and Totoaba are the only marine species currently on the list but there have been recent efforts to add the Atlantic Bluefin Tuna (*Thunnus thynnus*, Scombridae) to CITES.

Of the 124 endangered and threatened species protected in the United States, it is hard to say which are in greatest danger of extinction because many are on the brink. Fishes that live in springs and spring runs, especially in desert regions, and in caves are very sensitive to water pollution, water withdrawal, and introduced species. This group includes several minnows, pupfishes, cavefishes, darters, and topminnows. Other families with several endangered species are suckers and sturgeons. In the Pacific Northwest region, many distinct genetic races of salmon have gone extinct or are in need of protection.

Will fishes be affected by global warming?

Global warming is just one aspect of global climate change, and climate change will influence most of the environmental factors that determine where fishes live. These factors include temperature extremes, oxygen availability, floods, droughts, major storms, and habitat loss. All these aspects of climate change can potentially affect the genetics of populations, the interactions between fishes and plants and other animals, as well as the roles that fishes play within ecosystems.

Global warming will likely lead to alterations in wind direction and strength, which in turn will alter ocean currents. Ocean currents determine the timing and locales of spawning of ocean fishes, the dispersal of their larvae, and the distribution of adults. Altered currents could affect the distribution and production of open ocean species that make up 70% of the world's fisheries. In the North Atlantic, where bottom temperatures increased only 1°C between 1977 and 2001, 24 of 36 fish species moved northward or deeper toward cooler waters, indicating that shifts in species distributions have already occurred.

Increased temperature will influence freeze-thaw cycles, which determine spawning in many lake fishes. Northern Hemisphere lakes now freeze about 10 days later than they did 150 years ago, and snow is melting 1 to 2 months earlier than 50 years ago in parts of Europe. Earlier snow melt has reduced spring floods and disrupted fish migrations and spawning. Spawning migrations and timing of several freshwater fishes (pike, ruffe, bream, smelt) occur 12 to 28 days earlier than historical records indicate.

Increased temperatures are also a threat because fishes often live close to the maximum temperatures they can tolerate, and because oxygen concentrations in water are reduced at higher temperatures at the same time that metabolic rates increase. Also, many pollutants are more toxic at higher temperatures and are likely to become more concentrated due to drought. Corals are also very sensitive to slight increases in temperature, as was seen in 1998 when an only 1°C increase in average temperatures killed 50 to 100% of the corals in many areas of the tropical Pacific. Many coral reef fishes are directly and indirectly dependent on live corals, and these fishes declined in many regions.

A likely result of temperature increase that is already being detected is an increase in sea level, which will affect fishes that live near shorelines. Sea level rise will flood coastal wetlands, mangroves, and salt marshes. These habitats are major nursery grounds for numerous fish species and house the base of the food web for many other fishes.

As climates change, rainfall patterns will shift, making some regions wetter, others drier. Drought in combination with higher temperature will affect the distribution of many species as streams and ponds dry up and overheat. Cold-adapted species at high latitudes and altitudes will be pushed out of their habitats with no place to go because temperatures are rising faster than fish can evolve physiological traits that allow them to live at higher temperatures.

Many other changes have been predicted based on expected alterations to the world's climate. Some species will likely benefit from global warming, including warmwater species that can move farther north at northerly latitudes and coolwater species that can live at higher altitudes and latitudes. But such "gains" would be offset by an overall loss of genetic and species diversity, especially because climate appears to be changing too quickly for genetic change to keep pace. New species will not have time to evolve to take the place of those that cannot adapt. Overall, fish biodiversity will decline given current and projected change in the world's climate.

Are fishes affected by pollution?

Because fishes live in the water, absorbing it, breathing it, and swallowing it, they are more affected by pollution than land animals like us who only have to drink it. Fishes serve as indicators of the health of aquatic systems, and they can do so before human health problems arise.

Humans produce pollution in the form of toxic substances (chlorine, heavy metals, pesticides, herbicides, detergents, endocrine disrupting compounds, and oil spills), acid rain and snow, and silt and sediment. Fishes, both young and old, are made sick and die from some poisons, although the effects of some poisons are delayed while they work their way up the food chain. Some toxins also impair the development of young fish, slow sexual maturation, and even cause sex reversals. Human medications flushed down toilets or simply excreted in urine have been known to cause sex change in fishes. Male Smallmouth Bass in the Potomac River near Washington, D.C., have been found with eggs in their gonads, linked to a variety of pollutants in the water.

Acid rain and snow, caused mainly by air pollution from power plants, are particularly deadly for developing fishes and have wiped out many salmon and trout populations in North America and Europe. Silt and sediment carried in runoff from eroding hillsides settle on the bottom, killing eggs and covering up important habitat such as rocks and gravel. Silt in the water clogs gills, making breathing more difficult. Young fish raised in silty water suffer gill damage and grow more slowly.

Pollution can come in unexpected forms. In 2003, approximately 19,000 Paddlefish, catfishes, Freshwater Drum, sunfishes, and Largemouth Bass died in a Kentucky river when 800,000 gallons of bourbon were released from a warehouse after a fire. Fish died from alcohol poisoning and lack of oxygen, the oxygen having been used up by microorganisms that exploded in numbers after feeding on the alcohol.

 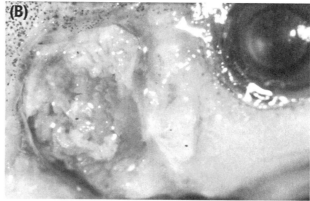

Silt in streams injures the gills of young fishes. Spotfin Chub (*Erimonax monachus*, Cyprinidae), a federally protected minnow, were raised in aquariums with different amounts of sediments. (*A*) Fish raised in clean water had healthy gills and fast growth. (*B*) Fish raised in very silty water suffered gill damage and grew more slowly. Photos courtesy of A. B. Sutherland

Why do people hunt and eat fish?

Fishes are caught for food, for sport, or to be kept as pets. Fishes are the main source of protein for perhaps one billion people, one-sixth of the world's human population. Another 2 billion people include fish as a part of their diet, meaning that half the world's human population relies on fish as an important protein source. Healthy fish populations are therefore important for human health, especially because many fishes are high in particular kinds of oils (omega-3 fatty acids) that are known to be good for people.

Commercial fishing is a huge industry that employs half a million people who catch 80 to 90 million tons of fish a year. The most important fisheries are for herrings and sardines, cods, pollock, anchovies, tunas, flounders, and salmons. In addition, another 40 million tons of fish are farmed in ponds and net pens. Species that are most commonly farmed are salmon, carp, tilapia, catfish, and trout. Individuals also catch fish to feed their families. Such subsistence fishing undoubtedly is done by many more people than sell their catch. Subsistence fishing involves hand gathering, spearing, netting, angling and trapping, and targets a larger variety of fishes than commercial fishing.

People keep fish as pets in home aquariums and ponds or watch them in public aquariums. Most aquarium fishes are tropical freshwater species raised in ponds, but some are captured in the wild. This is especially true for marine species. Some people catch fishes from local streams and lakes to keep as pets (see "Do fishes make good pets?" in chapter 8).

Fishing, both commercial and subsistence, while providing important protein, also has environmental costs. Poorly managed fisheries are overfished and many commercial species are in decline. Some commercial fish-

Two spearfishermen walk across a reef in Palau, Western Caroline Islands, for a day of fishing. Subsistence fishing of this type provides important protein for people in many tropical countries. Healthy reefs are crucial for such activities, and Palau has adopted strong reef protection laws, including a total ban on shark fishing.

ing methods such as dragging nets along the bottom (trawling) destroy habitats. Many non-targeted fishes, seabirds, marine mammals, invertebrates, and turtles are caught and killed during fishing. Dynamiting, which happens in coral reef fisheries, kills corals and many other reef organisms. Divers collecting reef fishes for the aquarium trade sometimes use poisons such as cyanide, which again kills corals, invertebrates, and many fishes that are not targeted. Also the species that are targeted and sold often die in peoples' fish tanks from delayed effects of cyanide.

Fish farming is controversial because of pollution from overfeeding and fish wastes, chemicals used to prevent diseases, fishes caught to feed cultured fishes (which reduces their availability as human food), contaminants in the flesh of cultured fishes, dyes used to color the flesh of cultured fishes, and ecological impacts of fishes that escape from aquaculture facilities.

Is there such a thing as fish leather?

Fish skin is often used to make leather because of the toughness and flexibility of fish skin when properly cured. The most common fish leather comes from salmons, eels, sharks, and hagfishes, but many other kinds can be tanned and turned into leather. Hagfish leather is usually sold as "eel skin," probably because calling it "hagfish leather" is a poor marketing tactic. But "eel skin" is soft, attractive, and sturdy, making excellent boots, belts, watchbands, briefcases, wallets, checkbooks, backpacks, and purses, some worth hundreds of dollars. True (anguillid) eel skin has been used historically in such products as buggy whips and door hinges, attesting to the quality and toughness of the material.

Fishes: The Animal Answer Guide

Colors are added to the feed of salmon grown in large pens to give their flesh a more natural, salmon-pink color. This bumper sticker is part of a campaign to encourage people to buy wild-caught, properly managed salmon, rather than farmed fish.

Why do so many fishes die at once?

Large numbers of dead fish floating in a lake or washed up on a shore are called "fish die-offs" or "fish kills." "Die-off" is the more general term, used when the cause is unknown or involves several, relatively natural factors such as low oxygen due to warm summer temperatures or thick ice cover (see "Can there be too many fish in a lake or river?" in chapter 9). A "fish kill" results from an identified cause involving a known, "responsible party," such as a toxic waste release, industrial accidents, ship groundings that lead to oil spills, or blooms of toxic algae caused by nutrients running off from poorly maintained agricultural land. A fish die-off is a matter for concern; a fish kill is a crime scene.

Are boats dangerous for fishes?

Little evidence exists to suggest that boats are a direct threat to most fishes; not many fish get chopped up by propellers or struck by a speeding hull. Fish are good at avoiding things moving quickly toward them, such as predators. One exception is sturgeons (Acipenseridae) that swim near the surface during the spawning season and suffer from boat and motor collisions. However, other things that boats do can have a direct or indirect, negative impact on fishes.

Problems result from boats operating at high speeds in shallow water. The wake from the boat and the turbulence from the propeller stir up bottom sediments, erode shorelines, cut up vegetation, and disturb fish nests. All of these problems can be avoided by requiring boats to slow down in areas less than 3 meters (about 10 feet) deep, and especially in water less than 2 meters (6 feet) deep.

Boats can cause fish injuries and deaths in other ways. Oil spills, either when people pump gas or when boats run aground, are a serious problem.

Human Problems

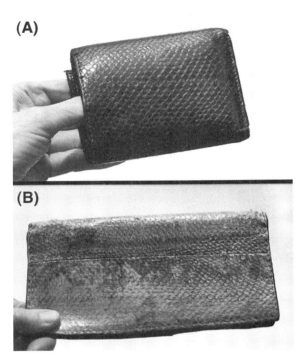

(A)

(B)

Fish skin is tough, flexible, and easily made into a variety of leather products, such as the wallet (A) and the checkbook (B) made from salmon skin.

Ship groundings on a large scale, such as freighters and tankers, cause massive damage to reefs and the fishes that live there, both because of fuel spills and from the grinding of the hull on the reef.

One mysterious series of fish kills in the Fox River of Wisconsin turned out to be a direct result of outboard motors, not boats. Over a 6 month period, 58 fish kills occurred, some involving more than 30,000 individuals. The cause turned out to be carbon monoxide moving downstream from a nearby outboard motor testing facility.

Outboard motors may have other, harmful effects on fishes. Croakers and seatrout (*Bairdiella*, *Cynoscion*; Sciaenidae) stopped making courtship calls when approached by outboard motor boats and did not resume their calls for up to a minute after the outboards were turned off. More dramatically, Fathead Minnows (*Pimephales promelas*, Cyprinidae) subjected for 2 hours to the sound of an idling 55-horsepower outboard motor lost the ability to hear sounds to which they are normally most sensitive. Those kinds of noises could commonly occur around fishes living near a marina or along a lake shoreline with heavy boat traffic.

How are fishes affected by litter?

Trash in a stream or on a lake bottom is still trash. Some trash leaks harmful chemicals that can affect fishes, as would any pollutant. Fish un-

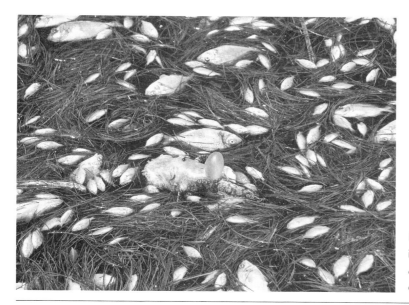

Fish killed by a toxic algae bloom in west-central Florida, 2005. Photo courtesy Florida Fish & Wildlife Conservation Commission

doubtedly consume some small plastic items that cannot be digested and that could block their intestinal tracts. A fish that had a plastic soda ring around its head would have trouble breathing because it could not open and close its gills. However, most of what we tend to think of as litter (beer cans and bottles, plastic soda rings, cigarette butts, plastic bags) has less impact on fishes than on many other aquatic organisms such as turtles, dolphins, and seabirds. Small fishes occasionally occupy cans and bottles on the bottom, at least in places where normal structure such as logs and rocks are absent.

The worst kind of litter, from a fish standpoint, is lost fishing gear such as old nets and traps. This "ghost gear" is usually made of indestructible material, usually monofilament or plastic, and continues to fish for weeks, months and even years, killing fishes by the thousands. Many state and federal agencies spend millions of dollars a year finding and removing ghost gear.

What can an ordinary citizen do to help fishes?

People can do many things to help fishes of all kinds. Buying and eating fishes that are caught and managed sustainably can reduce many of the problems associated with commercial fishing. This means buying species that are fished in a way that guarantees their success well into the future. Many campaigns have been successful in promoting sustainable practices and discouraging continued capture of overfished species such as Bluefin Tuna (*Thunnus thynnus*), Patagonian Toothfish (*Dissostichus eleginoides*),

Orange Roughy (*Hoplostethus atlanticus,* Trachichthyidae) are slow growing, slow to mature fishes that have been depleted in many southern ocean areas. Because they spawn in aggregations, they are easy to catch in large numbers. But because they do not mature until they are 25 to 30 years old, populations are replaced very slowly. Orange Roughy are not particularly good looking, which is why they are usually sold as filets.

Orange Roughy (*Hoplostethus atlanticus*), and many sharks, sturgeons, and groupers, to name a few. A good general source of information detailing which fishes to eat and not eat is the Seafood Watch program of the Monterey Bay Aquarium (www.montereybayaquarium.org).

For aquarium species, and to avoid the problems discussed in chapter 8 about fishes as pets, purchase fishes from sellers that have been certified by the Marine Aquarium Council (www.aquariumcouncil.org) and learn which species are caught in the wild versus bred in ponds. Although pond-reared fish usually have fewer negative impacts, some wild-caught fishes are harvested sustainably and support local collectors. These fishers would otherwise be forced into less ecologically friendly occupations such as cutting down rainforests. For information on sustainably caught freshwater species, see http://opefe.com/piaba.html. For marine species, buy only species on the American Marinelife Dealers Association's Ecolist of fishes that do well in home aquariums (www.amdareef.com).

We can all help protect the diversity of fishes locally and worldwide by reducing our need for fossil fuels (petroleum products, coal) and all the problems that go along with their mining, transport, processing, burning, and waste disposal. Buying other "green" products whenever possible pro-

Fishes: The Animal Answer Guide

tects habitats in which fishes and other organisms live. Educating yourself about habitat issues, pollution problems, overfishing, introduced species, energy production, unwise aquaculture practices such as salmon farming, and climate change will make the world a better place to live, for fishes and for humans.

Chapter 11

Fishes in Stories and Literature

What roles do fishes play in religion and mythology?

The sea is a dangerous and mysterious place and understandably the source of many myths. Innumerable sea monsters gobbled up sailors, landing a place in history. Some of these monsters were in reality whales, giant squid, and sharks, others were bony fishes, and some were the result of imaginations fed by the perils of long ocean voyages, without decent photographic equipment. But their place in mythology and religious ceremony is well established around the world, as in the ancient Babylonians, Assyrians, Phoenicians, and Philistines, all who worshiped a half-fish, half-human god.

An apparent source of many sea serpent stories, particularly those describing a giant beast "having the head of a horse with a flaming red mane," were Oarfish (*Regalecus glesne*, Regalecidae). Oarfish are probably the longest bony fish in the world, reaching a reported length of 11 meters (35 feet). They have a bluish-silvery body, scarlet head crest with long fin rays, and deep red fins. It is unlikely that one ever ate a sailor, since Oarfish feed primarily on tiny fish and invertebrates. Other sitings of sea serpents are now thought to have involved decomposing Basking Sharks (*Cetorhinus maximus*, Cetorhinidae), which grow to 12.5 meters (40 feet) long and leave behind a cartilage skeleton with a huge head and a long tail.

Long sea voyages also spawned many tales of mermaids, creatures with the upper body of a woman and a lower half of a fish (that few mermen appear in myths probably reflects the fact that ships were staffed by lonely men). Most mermaid reports were likely based on sightings of dugongs or manatees, which are marine mammals with whalelike tails, a recognizable face (but with whiskers), and that nurse their young as any good mammal would. Some documented accounts included drawings. The eighteenth-century Dutch artist Samuel Fallours drew a 59-inch mermaid that he said he kept in a bathtub in Indonesia for 4 days and 7 hours. The drawing was published along with other fanciful sea creatures by natural historian Louis Renard. Mermaids appear in the legends of non-sailors too. A *Jengu* is a creature of ancient Cameroon, Africa, lore who is much like a mermaid that lives in rivers and the sea and brings good luck to those who worship her.

Many large lakes house monsters and serpents that may (or may not) be fishes. The best known is Nessie of Loch Ness, Scotland. But Nessie is far from alone, although she (?) is separated by continents and oceans from others of her apparent kind. Lake Erie has South Bay Bessie, a 10 to 12 meter (30 to 40 foot) grayish creature that is said to resemble a snake or an eel. Crescent Lake, Newfoundland, has Cressie. Lake Tahoe, California is home to Tessie. Canada's Lake Okanagan has Ogopogo, Lake Champlain of New York and Vermont has Champ. Still unnamed is a 10 meter (30 foot) monster in the White River of Arkansas that some people think is actually one or several sturgeons or Paddlefish. Alkali Lake in Nebraska has a 12-meter (40 foot) creature with a horn between its eyes. In Alaska, Illiamna Lake's monsters are grayish, broad headed, with bodies that are 20 meter (60 foot) long and vertical tails. Utah has at least five different monsters in five different lakes. Lake Tianchi, China, also has a monster.

Are these fishes from a bygone era, or plesiosaurs, or logs, or ripples in afternoon sunlight? Who's to say?

Other large, feared fishes are unquestionably real. Sawfishes (Pristidae) live in nearshore, tropical areas and often enter fresh water. They occur, or occurred, in the river mouths and bays of Australia, Florida, Central America, and in tropical West Africa. These huge, powerful relatives of skates and rays can reach lengths exceeding 7 meters (23 foot) and weights of over 5,000 pounds (2,270 kilograms). Projecting from either side of the sword-like snout are rows of 1-inch long, narrow, spiky teeth that are used for slashing sideways through schools of prey. A blow from a large sawfish can cut off a person's arm or foot.

Aztecs treated sawfish as an Earth monster, and Asian tribal doctors used the snout to chase away demons and cure disease. Among coastal African societies, sawfish are considered among the most dangerous beasts one can encounter, with supernatural powers. Fishers can harness these powers to help their community, the sword having the ability to repel evil forces, sickness, and misfortune. The Duala of Cameroon have proverbs about sawfish; the Lebou of Senegal adorn their houses with saws to protect their homes, family, and cattle; the Jola of Senegal and Gambia regard sawfish as

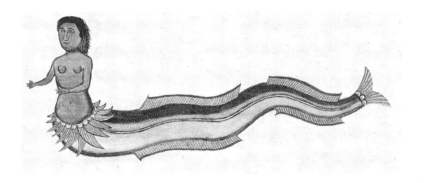

Drawing made around 1710 by Samuel Fallours of a 59-inch mermaid that he said he kept in a bathtub in Indonesia. Fallours claimed that he lifted its fins "in front and in back and [found] it was shaped like a woman." In the original, the fish part of the body is blue, the fins are green edged in red, the human body is greenish, eyes are blue, and hair is brown. Drawing reproduced in Pietsch 2010

totemic ancestors and the sword as a magical weapon; the Akam of Ghana treat the sawfish as a symbol of authority.

Myths and legends involving fishes occur in many cultures. An Australian Aboriginal tale explains that fishes were at first land animals vulnerable to the cold. They find that they are always warm in the water, as plausible an explanation for cold-bloodedness as many others. Another Aboriginal tale describes how Barramundi (*Lates calcarifer*, Latidae) were created when two young lovers, Yungi and Meyalk, from different tribes (another recurring theme) hurled themselves into the water rather than be separated. They became Barramundi, and the sharp spines in their dorsal fins are the remains of spears hurled at them by pursuing tribesmen. In a Hindu version of the Great Flood, Manu, the first human, saves a small fish that grows into a giant fish called a Ghasha (possibly a Golden Mahseer, *Tor putitora*, Cyprinidae), which warns Manu of the upcoming flood and eventually pulls the ark to safety.

Many myths tell of giant fishes responsible for the creation of nations and the heavens, that help hold up the world, or that are responsible for earthquakes, tidal waves, erosion, and strong winds. A New Zealand Maori legend recounts how Maui caught a giant fish that eventually became the South Island. Maori legend also tells of Ikaroa, also known as Mangoroa, a long fish (perhaps a shark) that gave birth to the stars of the Milky Way. Bahamut is a giant fish that supports the earth in Arabian mythology, the earth resting on a mountain that rests on a bull that sits on sand that rests on the fish. Aztec mythology recounts Cipactli, a part fish (plus toad plus crocodile) sea monster of infinite appetite. The gods created the Earth from Cipactli's body.

A Chamorro legend from the island of Guam recalls a monster fish that was eating its way through the island every night. A group of girls wove a

Fishes: The Animal Answer Guide

Sawfishes are worshipped around the world because of their size and strength. Regrettably, sawfishes were once common but are now among the most threatened fishes in the world because they become entangled in nets of all types. In the United States, sawfishes occur today only in southern Florida, primarily in the Everglades region. Shown here are results from a fishing tournament around 1920, in Key West, Florida. Some of the fish were said to have weighed 1,700 pounds. Photo property of Matthew McDavitt

huge net from their hair and began to sing. The monster fish was attracted by their singing, and the girls caught the monster and saved the island.

Japanese mythology tells of a giant catfish, Namazu, that is responsible for earthquakes. Normally, Namazu is restrained by the god Kashima but when Kashima lowers his guard, Namazu thrashes about, making the earth shake. In east African legends, a giant fish carries a stone on his back and a cow stands on the stone, balancing the Earth on one of her horns. When her neck aches from the weight, she tosses the Earth from one horn to the other, causing earthquakes. In the Solomon Islands, tidal waves are caused by the thrashing of a giant lake eel, Abaia, who reacts violently to anyone foolish enough to catch fish from the lake, all of which are her children.

Another Japanese giant fish is Isonade, a sharklike sea monster that causes strong winds and also capsizes boats or catches unwary fishermen with the hooks on its huge tail.

What roles do fishes play in Western religions?

Several of Jesus' disciples were former fishermen on the freshwater Sea of Galilee, before and after Jesus rewarded Simon for loan of a boat by filling his net and the nets of others with fishes (Luke 5:1–11). Peter caught a fish with a gold coin in its mouth that Jesus used to pay his temple tax (Matthew 17:24–27). The fish is now known as St. Peter's fish and is probably the tilapia *Sarotherodon galilaeus galilaeus*. Another story explains two spots on either side of the fish, caused by Peter--formerly a charcoal

Fishes in Stories and Literature

maker—picking it up gingerly with two fingers and tossing it back because it was so ugly. This story more likely applies to another fish, the John Dory or St. Pierre Fish (*Zeus faber*, Zeidae), which actually does have a black spot on either side of its body and is, to many people, ugly. The cichlid does not have two black spots, nor is it strikingly bad-looking.

In the miracle of the seven loaves and fishes (Mark 8:1–9), Jesus fed thousands with seven loaves of bread and a few small fish. In Matthew 13:47–50, Jesus likens God's decision concerning who will go to heaven and who to hell to fishermen sorting their catch, keeping the good fish and tossing the bad fish, an action known today as "high-grading." And Jesus referred to his disciples as "fishers of men." "Ichthys" (the root of ichthyology) is the classical Greek word for fish, the letters standing for the first letters of five words referring to "Jesus Christ, God's son, savior." A corresponding figure is made of two intersecting arcs that cross at one end to resemble a fish. It was used by early Christians as a secret symbol and has become known as the "sign of the fish" or the "Jesus fish." The Jesus fish has become a symbol of modern Christianity, depicted on bumper stickers and jewelry to show that the owner is a Christian. The symbol has spawned a variety of spin-offs, and some argument, including the Darwin Fish, the gefilte fish, fish and chips, lutefisk, and Jesus fish eating Darwin fish.

In the Old Testament, Jonah was swallowed by a great fish (Jonah 1:17). However, most modern translations depict the great fish as a whale. The biblical story of Jonah also appears in the Koran. The story has also taken on a popular meaning, where a person referred to as "a Jonah" is someone who brings bad luck to a venture, especially a sea venture.

Biblical restrictions on eating seafood, that is, kosher dietary rules, date back to Moses restricting the Israelites' diet to sea creatures that have fins and scales (Leviticus, xi). This rules out eels and catfish, among others. Fish were apparently a ritual part of Sabbath meals among the early Hebrews, and the Jewish Encyclopedia says that fish should always be present at the Sabbath meal no matter how difficult it is to get. In the Talmud, fish are linked to righteousness, fertility, social harmony, protection from evil, and contentedness.

The Catholic requirement to eat fish on Fridays relates to a tradition from the early days of Christianity that required abstinence from eating other kinds of flesh on numerous days. Slowly the other days were eliminated but the Friday tradition held. Fish remain a strong symbol in Catholicism, the pope's traditional costume including a fish-shaped hat (the first to hold the office was the apostle Peter, who was also a fisherman before he was pope).

Hindus also revere fishes, and Hindu gods include two important fish. Matsya, a fish, is the first incarnation of the Lord Vishnu, and Ganga, the

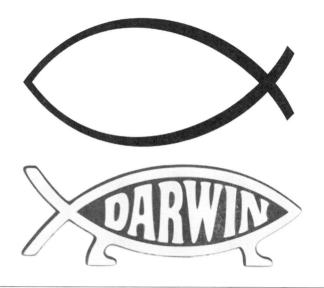

Variations on a religious theme. (*Top*) Early Christians identified themselves with an image of two intersecting arcs that cross at one end to resemble a fish. This symbol is popularly known as the "Jesus fish." (*Bottom*) Humorous as well as contentious variations include the "Darwin fish."

goddess of the Ganges River, rides upon a half crocodile, half fish. The bride in a traditional Bengali wedding carries a fish in her hand, which is a symbol of social harmony, fertility, and sexuality. In Buddhism, the Buddha is often shown as having eyes made of golden fishes with which he gazes upon the world with compassion.

Did early philosophers and naturalists mention fishes in their writings?

Aristotle (384–322 BC) is considered by many as the Father of Zoology. He was the first (among Western philosophers) to make careful observations and dissections, at the time a revolutionary act. He classified all living things, including fishes. He was also the first to realize that some sharks produce eggs that hatch within the mother; published detailed descriptions of the anatomy of catfish, electric torpedo rays, and anglerfish; and was the first to explain how parts of the fish stomach called pyloric caecae functioned in digestion.

Inevitably, Aristotle got a few things wrong. In his dissections of European Eels (*Anguilla anguilla*, Anguillidae), he never found a female with eggs and concluded that they arose "spontaneously" from muddy lake bottoms, perhaps as growths from horse hairs. His misunderstanding can be forgiven since it was not until the 1800s that researchers worked out the details of eel reproduction, which involves a spawning migration that takes eels from Europe to a western tropical Atlantic region known as the Sargasso Sea. Females don't produce recognizable eggs until they are well into the oceanic part of their migration.

Aristotle's dissections of torpedo rays were relevant because these marine electric fishes, which produce strong electric shocks equivalent to about 1 kilowatt of power, were used as shock therapy to cure a variety of ailments, including epilepsy.

Students of Aristotle explained more fish biology. One topic that fascinated a number of early natural historians, especially Theophrastus (371–287 BC) was overland movement of some Egyptian fishes when rivers dried up (probably clariid "walking" catfish), and the discovery that some fishes laid eggs in the bottom of ponds that could survive drought, hatching with the next rains (probably cyprinodont killifishes).

Even before Aristotle, philosophers were speculating about where fishes fit in the great scheme of things. Anaximander (611–547 BC), another Greek philosopher, wrote that fishes were the first animals, arising "from the original humidity of the earth." Fish then came ashore, lost their scales, and were the ancestors of all land creatures including humans. His ideas are not far from the truth and Anaximander is recognized as the first evolutionary biologist, 2,400 years before Charles Darwin.

The Romans were active fish farmers and are credited with keeping the first Carp in ponds. Pliny the Elder (AD 23–79) was a Roman natural historian and wrote one of the first and most complete early volumes on natural history, *Naturalis Historia*. In it he discusses fish farming, aquarium keeping, and that aquarists of the times were too fond of the fishes in their "glass boxes."

Are fishes in fairy tales?

If you accept merpeople as fish, or at least part fish, then "The Little Mermaid" (by Hans Christian Anderson, not Disney) would have to count as a popular fish fairy tale. Stories of girls turned into fish turned into girls, and vice versa, occur in many forms from many places. Inevitably, romantic interest in another species drives the tale. The 2009 animated feature, *Ponyo*, by acclaimed Japanese filmmaker Hayao Miyazaki depicts the wannabe human as a Goldfish (*Carassius auratus*, Cyprinidae). Miyazaki endows his Goldfish with magical powers, including an ability to live in salt water. Grimms' Fairy Tales include "The Fisherman's Wife," in which an old, poor fisherman catches a magical fish (in some versions it is golden, in others it is a flounder). He releases it but his wife feels the fish owes them a favor or two. The wife's greed and the favors escalate one upon another until finally the cycle is completed and the couple winds up where they started.

Fishes play limited but pivotal roles in many other fairy tales. In "The Fish and the Ring," a large fish (species undetermined) swallows a ring that reunites a young nobleman and his much-abused wife. Unfortunately, the fish becomes a meal. In versions of the story of "Tom Thumb," Tom is swallowed by a fish that is served to the king (a repetition of the fish-as-dinner theme), at which time Tom emerges when the fish is being carved up, Tom having somehow survived being cooked to death. In T. H. White's *The Once and Future King* (animated in Disney's *The Sword and the Stone*), Merlin turns the future King Arthur into a minnow (Cyprinidae) that barely escapes being eaten by a Northern Pike (*Esox lucius*, Esocidae).

What is *gyotaku*?

Gyotaku is a modern Japanese art form for making accurate, life-size fish prints. *Gyo* means "fish" and *taku* means "rubbing." Rather than trying to draw the fish, the artist rolls the fish in ink and then presses it onto paper. At its simplest, *gyotaku* involves little more than making such a rubbing, and anyone can do it. We both have had our students make *gyotaku* prints on paper and on t-shirts as a class exercise. Making *gyotaku* prints has become a popular classroom exercise for students of all ages and helps people appreciate the beauty of nature. For anglers it is an interesting way to "preserve" one's catch.

The artistic part of *gyotaku* involves filling in the details, such as color, eyeballs, creating depth, putting multiple fishes on a page, adding seaweed or other surrounding objects. Some skilled *gyotaku* artists sell their finished prints for large sums.

What roles do fishes play in various cultures?

It is no surprise that fishes have played an important part in many cultures. Fishes are a readily available source of protein that can (or previously could) be caught year around and in large numbers. In addition, fishes were used to make medicines (seahorses), fertilizers, jewelry, needles for sewing and tattooing (various bones), weapons (shark's teeth swords), leather (eels, salmon), were boiled down for oils or burned as lamps when dried (candlefish), their scales were used in fishing lures (parrotfish), and bones and skin were boiled to make gelatin. Salmon, because they enter rivers to spawn each year in great numbers, have been central to the lives of many native peoples. The complex cultures and extensive art of many tribes in the U.S. Pacific Northwest, Canada, and Alaska were in part a result of the abundance of Pacific salmon species (*Oncorhynchus*, Salmonidae). This richness left people time for activities beyond pure survival. Many tribes had annual, elaborate "First Fish" ceremonies, celebrating the start of the spawning run. The first salmon caught each year was honored and even returned alive to the sea to insure a successful harvest. The Klamath tribes of the Klamath River region of southern Oregon and northern California depended on spawning runs of lake suckers (Catostomidae) and had similar first fish celebrations to honor the event.

Salmon in large numbers led to folklore in other regions where these fishes occurred. Atlantic Salmon (*Salmo salar*) were abundant throughout western Europe and played important roles in regional mythology. Fionn mac Cumhaill was a mythological hunter-warrior in Irish mythology who was spattered with a drop of fat while cooking "The Salmon of Wisdom." Upon sucking his burnt thumb and swallowing the drop of fat, he acquired all the knowledge of the world, with which he became the leader of the Fianna, the heroes of Irish myth. In Welsh mythology, the poet Taliesin received his wisdom by similar means.

Throughout the tropical Pacific, island cultures developed that were dependent on the sea, with much local wisdom and lore focused on sea creatures and their relationships to humans. For example, in the Palau Islands of Micronesia (Republic of Belau), legends tell of how butterflyfishes got their two black spots (similar to the story of the St. Pierre fish and its spots, explained earlier, except involving Palauan gods, not the apostle Peter), why trunkfish have two horns pointing forward and two back, how parrotfish got slimy teeth, why the waves are biggest when the mullet are spawning, what it means when a shark bites the bottom of your canoe, and numerous stories about lovers involved with moray eels, turtles, dugongs, and other sea creatures. Many of these stories are carved into the walls of the men's meeting houses or onto storyboards, now sold to tourists.

Stories about gods being involved in coloring fishes occur in many cultures. For example, in Hawaiian legends, Kapuhili (the Raccoon Butterflyfish, *Chaetodon lunula*) was responsible for the initial marking of red, yellow, and white fishes; the remainder he merely spotted with ashes. Kapuhili then gave them all their names. As in Palauan, and other cultures, the behavior of fish has been interpreted to explain the behavior of humans. In Palau, when the shark swims upside down and bites your canoe, it means your wife is cheating on you. In Hawaii, similar unfaithful behavior at home was indicated when parrotfishes were seen rubbing noses. Apparently, men at sea spent much time worrying about what their wives were doing while they were away.

What roles do fishes play in popular culture?

Fishes show up everywhere, in books, cartoons, movies, music, TV shows, postage stamps, t-shirts (we have dozens, some of the best found at www.trollart.com), and so on, attesting to their popularity with the public.

On TV, excellent documentaries about fishes, the ocean, and fresh waters

Salmon figure prominently in the art of many native tribes along the west coast of North America. Many of the typical art symbols and designs are evident in this cedar carving of salmon on a spawning run, including the use of ovals in the eyes, split Us in the tails, and tapered Ls in the fins. Inset, upper right: details of the head of the fish in the upper left of the school Carving by Ron Aleck of the Pelakeet Band, British Columbia

abound. These usually accurate depictions of nature include specials from the BBC (the *Planet Earth*, *Blue Planet*, and *Life* series in particular), National Geographic, NOVA, Nature, IMAX, Disney, Animal Planet, and the Discovery Channel, to name just a few. The Discovery Channel gets mixed reactions from shark lovers with its annual, immensely popular "Shark Week," which seems designed to frighten as much as enlighten. In the category of TV cartoon series, fishes get a bum rap in *SpongeBob SquarePants*. Flats the Flounder is a bully, Bubble Bass is generally hated, Tom the Bartender gets few speaking lines, and Scooter the Surfer Dude fairs only slightly better until SpongeBob kills him. So what does Stephen Hillenburg have against fish?

Some popular bands have had fishy names. Country Joe and the Fish was big in the 1960s (watch the movie *Woodstock* and learn the cheer). Eels is an alternative rock band that started in the mid-1990s and whose music has been featured in all three *Shrek* movies. The jam band Phish had a huge following in the 1990s, broke up, and reformed in 2009. Among ichthyologists, a popular band that sings about fishes is The Ratfish Wranglers, out of Ketchikan, Alaska.

Hollywood has featured fishes prominently or in bit parts in a number of movies. Some great lines have involved fishes: "Don't eat the green ones, they're not ripe yet" (*A Fish Called Wanda*), "Don't eat the fish" (*Airplane!*), "I caught you a delicious bass" (*Napoleon Dynamite*), "Beneath this glassy surface, a world of gliding monsters!" (*Deep Blue Sea*), and of course "You're gonna need a bigger boat." (*Jaws*). The *Jaws* theme may be the best-known two-note musical phrase in the world.

Sharks play pivotal and often villainous roles in many movies. In *The Old Man and the Sea* sharks tear apart the giant Blue Marlin (*Makaira nigricans*, Istiophoridae) that Santiago finally catches after 84 days without a

A Palauan storyboard relates a legend, with several actions depicted in a single scene instead of the multiple panels of modern cartoons. This board tells of a fisherman whose canoe is bitten by a shark. Knowing such activity means his wife is cheating on him, he returns to the village and spears his wife's lover.

fish. *Open Water* probably set the diving industry back a decade and ended unhappily as a result of shark activities. In *Deep Blue Sea*, genetically engineered Mako Sharks (*Isurus*, Isuridae) threaten the world but meet an untimely end in several impressive explosions. In *The Deep*, sharks and the giant Green Moray Eel (*Gymnothorax funebris*, Muraenidae) are the nonhuman villains.

Some other memorable fish-themed or fish-focused movies include *Piranha*, which involved more genetically engineered fish as part of "Operation Razorteeth," an effort to produce a mutant strain of piranha to use in the Vietnam War. The original movie was followed by a sequel in 1981, with another in 2010, this time in 3-D. *The Perfect Storm* contained some of the best swordfishing scenes ever filmed, including bit performances from Blue and Mako sharks (and also some bad weather). *The Life Aquatic with Steve Zissou* is a favorite among ichthyologists. Bill Murray as oceanographer Steve Zissou at one point complains that he needs his estranged wife along because he can never remember the *\$&@!#* scientific names of fishes.

Several movies had brief but memorable appearances by fishes. The first *Free Willy* movie included a scene at Pike Place Fish Market in Seattle where the (human) hero intercepts a thrown Chinook Salmon (*Oncorhynchus tshawytscha*, Salmonidae) to get a real meal deal for Willy. Some very fanciful fishes, some that looked like members of extinct Devonian groups, appeared in *Star Wars Episode I: The Phantom Menace*, the first Star Wars movie featuring Jar Jar Binks, perhaps the most hated character in filmdom. One of the most memorable brief fish appearances occurred in *The Naked Gun: From the Files of Police Squad!* Anyone familiar with just

Fishes: The Animal Answer Guide

how toxic and painful lionfish (*Pterois*, Scorpaenidae) spines are squirmed as Leslie Nielsen's character tried repeatedly to shake one off his hand.

Many movie titles lured us into thinking they were about fish, but in reality they were just teasing (*Big Fish, Little Fish, Rumble Fish, Passion Fish, Go Fish, Dead Fish, Albert Fish, Fish Tank, The Day the Fish Came Out*).

The best-animated film about fishes ever made has to be *Finding Nemo*. Recognizable reef species play all the important roles and the biology is correct, with one minor exception (see "How do fishes reproduce?" in chapter 6 about sex change in anemonefish). *Shark Tale* gave us kinder, gentler sharks, although many had split personalities and were in need of therapy to deal with their aggressive tendencies. More fanciful was *Ponyo*, about a goldfish that wanted to be a girl. In *The Little Mermaid*, Ariel's sidekick, Flounder, was anything but, and moray eels were definitely unlovable.

Fishes also appear in other forms of popular culture. Some comic strips have featured fishes and even ichthyologists. *Sherman's Lagoon* stars a shark, and one of the main characters in *B.C.* is Clumsy Carp, an avid and dedicated ichthyologist and fish watcher. In video games, well-known but apparently not very well-liked is Magikarp, a Pokémon (Pokemon character) that looks like a cross between a goldfish and a catfish with a nunchuk instead of a dorsal fin. Magikarp has tremendous leaping ability and an immune system capable of handling the most polluted waters. The videogame *The Legend of Zelda--Twilight Princess* contains the character Reekfish.

Finally, English is full of sayings and proverbs that revolve around fish: "A fish out of water," "there are plenty of fish in the sea," "a woman needs a man like a fish needs a bicycle," "she's a cold fish," "a grip like shaking hands with a dead fish," "like shooting fish in a barrel," "fishing for compliments," "fish or cut bait," "going to sleep with the fishes," "having bigger fish to fry," "that's a fine kettle of fish," "neither fish nor fowl," "a big fish in a small pond," "he drinks like a fish," and "a bad day of fishing is still better than a good day of school."

What roles have fishes played in poetry and other literature?

American poets have used fishes as subject matter, describing them directly or for comparisons of the human condition. Carl Sandburg, in his poem, "Flying Fish," describes a flying fish viewed from a boat as, "Child of water, child of air, fin thing and wing thing." Sandburg goes on to say that he, too, is often caught in the middle of some dilemma. A popular children's poem by Meish Goldish ponders the seemingly carefree existence of fishes, and its hidden dangers:

How I wish
I were a fish!
My day would begin
Flapping my fins.
I'd make a commotion
Out in the ocean.
It would be cool
To swim in a school.
In the sea,
I'd move so free,
With just one thought:
Don't get caught!

One of the most famous American poems containing a story about fishes is "The Song of Hiawatha," by Henry Wordsworth Longfellow. The character Hiawatha is based on the legendary leader of the Iroquois confederacy of Native American Indians. The poem mentions numerous fishes, and in Chapter 8, "Hiawatha's Fishing," Hiawatha goes out in his birchbark canoe using cedar bark line "to catch the sturgeon Nahma, Mishe-Nahma, King of Fishes." The sturgeon, undoubtedly a Lake Sturgeon (*Acipenser fulvescens*) given the Lake Superior setting that Longfellow adopts, is portrayed accurately ("There he lay in all his armor; On each side a shield to guard him, Plates of bone upon his forehead, Down his sides and back and shoulder, Plates of bone with spines projecting"). Hiawatha passes up other fish (Sahwa, the Yellow Perch; Maskenozha, the Northern Pike; Ugudwash, the sunfish) until finally the Great Sturgeon in a rage rises up and swallows Hiawatha, canoe and all. Hiawatha crawls about and finds the sturgeon's heart and lands a blow that kills the sturgeon, which floats to the shore where Hiawatha is freed by the scavenging activities of seagulls. The sturgeon is so large that it takes 3 days to boil its flesh into useful oil.

Among English poets, William Shakespeare's writings are full of fishes and fishing. Scholars have found about 200 such references in Shakespeare's writings, mostly involving freshwater fishes such as salmon, trout, pike, dace, carp, tench, loach, gudgeon, eels and minnows, with some mention of marine species such as mackerel and herring. Fishes are presented as food and as bait, but more often fish and fishing are associated with human traits, conditions, and behavior. In *Pericles*, the likelihood that the poor will find justice is described: "Here's a fish hangs in the net like a poor man's right in the law. 'Twill hardly come out." On making wise decisions, Iago, in *Othello*, says that Desdemona was not so foolish as "To change the cod's head for the salmon's tail," a cod's head being much more desirable food than the salmon's tail. In *Henry IV, Part II*, an elderly Falstaff ridicules the younger Justice Robert Shallow with, "If the young dace be a bait for the old pike, I see no reason in the law of nature but I may snap at him." Carp were considered a particularly difficult fish to catch and in *All's Well That Ends Well* a clown describes a rogue as being as cunning as a carp. In contrast, gudgeons, another type of minnow, were easy to catch and Shakespeare referred to simpleminded people as gudgeons.

Fishing is used repeatedly in Shakespeare's plays, often in the context of fooling people, as in *Much Ado About Nothing*, when Claudio, about to trick Don Pedro, says, "Bait the hook well. This fish will bite." Poaching trout (probably a crime punishable by death given that all trout belonged to the Crown) could be done slyly and quickly by wading along a stream, reaching under the bank, and tickling a trout under its head. In *Twelfth Night* the roguish Malvolio is called "the trout that must be caught with tickling." Scholars think Shakespeare was a trout poacher as a boy. Shakespeare describes the ironies of the food chain in the famous grave digging scene in Hamlet: "A man may fish with the worm that hath eat of a king, and eat of the fish that hath fed of that worm."

One of the most vivid fishing scenes, which of course has multiple meanings, occurs in *Antony and Cleopatra*, when the queen tells her maid:

Give me mine angle [fishing equipment]. We'll to th' river. There,
My music playing far off, I will betray
Tawny-skinned fishes; my bended hook shall pierce
Their slimy jaws, and as I draw them up
I'll think every one of them an Antony,
And say, "Ah ha, you're caught!"

In his poem "Mandalay", Rudyard Kipling wrote:

"Come you back, you British soldier; come you back to Mandalay!"
 Come you back to Mandalay,
 Where the old Flotilla lay:
 Can't you 'ear their paddles chunkin' from Rangoon to Mandalay?
 On the road to Mandalay,
 Where the flyin'-fishes play,
 An' the dawn comes up like thunder outer China 'crost the Bay!

However, the "road" from Rangoon to Mandalay lies along the Irrawaddy River and unfortunately for the accuracy of the poem, flyingfishes (Exocoetidae) are marine and do not live in rivers (see "Can any fishes fly?" in chapter 2)

Fishes also figure prominently in recent children's books. Dr. Seuss created a variety of fantastic fishes in *McElligot's Pool* and again in *One Fish, Two Fish, Red Fish, Blue Fish*. A pet fish played the voice of caution in *The*

Cat in the Hat. Leo Lionni's *Swimmy* portrays a tuna as the predatory villain, and in his *Fish Is Fish* we find out what happens when fish attempt to go amphibious. In Helen Palmer and P. D. Eastman's *A Fish out of Water* we all learned the dangers of overfeeding Goldfish more than a pinch, as Otto grew to ginormous proportions. Joanna Cole and Bruce Degen created The Magic School Bus, with Ms. Frizzle taking us on fish-themed field trips in *The Magic School Bus on the Ocean Floor*, *The Great Shark Escape*, *The Magic School Bus Takes a Dive*, and *The Magic School Bus Goes Upstream*.

Apart from the hundreds (or thousands) of books about how to fish are many famous novels, too numerous to list. Zane Grey and Ernest Hemingway were both avid fishermen. Grey wrote about fishing adventures in many books, in addition to his well-known novels about cowboys and western life. Hemingway described heroic scenes of big game fishing in his novels, including swordfishing in *Islands in the Stream*, and marlin fishing in his best known "fishing novel," *The Old Man and the Sea*. Flyfishing for trout was the means of keeping a family together in Norman Maclean's *A River Runs through It*. Some of the less admirable aspects of bass fishing tournaments highlight Florida author Carl Hiaasen's *Double Whammy*.

Five recent, well-written nonfiction books about single species are *Cod: A Biography of the Fish that Changed the World* (about Atlantic Cod, *Gadus morhua*, Gadidae) by Mark Kurlansky; *Blues* (about Bluefish, *Pomatomus saltatrix*, Pomatomidae) by John Hersey; *The Founding Fish* (about American Shad, *Alosa sapidissima*, Clupeidae) by John McPhee, *The Most Important Fish in the Sea* (about menhaden, *Brevoortia*, Clupeidae) by H. Bruce Franklin, and *An Entirely Synthetic Fish: How Rainbow Trout Beguiled America and Overran the World*, by Anders Halverson. The category of "how to" books began in the seventeenth century with Isaak Walton's *The Compleat Angler*, at the time a revolutionary volume that went against Puritan prohibitions against people engaging in sports in groups.

Ian Fleming, author of the James Bond 007 novels, wrote a short story called "The Hildebrand Rarity" about collecting fishes for the Smithsonian Institution (academic home of the second author of this book). Bond travels on a yacht to the Indian Ocean to help collect a rare, spiny, pink and black fish (apparently a scorpionfish) called The Hildebrand Rarity. The collector/villain, Milton Krest, beats his wife with a stingray tail (called "The Corrector") and in turn winds up dead, having had a Rarity shoved down his throat while very drunk. No one mourns the loss of Krest nor, unfortunately, the fish.

Finally, a few authors have managed to capture the beauty of fishes in a single phrase. One that stands out in recent literature occurs in an award-winning novel by Yann Martel called *Life of Pi*. The main character, Pi Patel, survives the sinking of a zookeeper's ship and winds up in a lifeboat with a tiger, who fortunately suffers from seasickness and therefore does not eat Pi. Pi survives by catching turtles and fishes, including Dolphin-fish, also known as Mahi-mahi or *Dorado* because of their golden coloration (*Coryphaena hippurus*, Coryphaenidae). Mahi-mahi are also well-known for their color-changing habits when hooked and brought on board a boat. Pi catches a Mahi-mahi and kills it with a club, which he describes as "beating a rainbow to death."

Do fishes have culture?

From all the information in this chapter, it is obvious fishes play a large part in cultural traditions and practices. But culture is not confined to the human species. Culture involves passing important social information such as traditions from one generation to the next. It occurs through teaching and learning, as when young individuals observe and repeat the behavior of older individuals. Biologists have found cultural activities in a number of nonhuman animals, such as young chimpanzees observing older chimps using a grass stem to "fish" termites out of a termite mound, and adult oystercatchers teaching young birds how to open clams and oysters.

Traditional behaviors and activities are also passed across generations of fishes. Numerous reef fish species reuse traditional breeding locales while perfectly adequate, nearby sites are ignored. Bluehead Wrasses (*Thalassoma bifasciatum*, Labridae) reuse the same breeding territories for more than 12 years, although individual blueheads seldom live for more than 3 years. Juvenile grunts (*Haemulon*, Haemulidae) rest in small schools over particular coral heads by day and migrate out into grassbeds at twilight to feed on invertebrates. The same coral heads and migration routes are reused for more than 3 years, although no fish is more than 2 years old, and while other coral heads and potential migration corridors are ignored. Young fish settle out of the plankton, take up residence where they find older fish, and then follow older fish each evening and morning between coral reef and grassbed, learning the sites and routes. The social traditions of daytime resting site locale and twilight migration route are thus established and maintained via cultural transmission.

Chapter 12

"Fishology"

Who studies fishes?

People who study fishes are called ichthyologists (derived from the Greek *ichthos*, "fish," and *-ology*, "study of"). They figure out how to differentiate between different species of fishes and work out their evolutionary relationships. They study ecology, behavior, physiology, genetics, and distribution of these different species of fishes to answer many of the questions addressed in the first 11 chapters of this book. Fishery biologists use this information plus additional information on life histories and population sizes to manage fish populations so that we can continue to harvest fishes for food and sport without endangering the survival of the species.

Which species of fishes are best known?

The best-known species of fishes fall into four categories important to humans: food fishes, aquaculture fishes, sport fishes, and pet fishes. We try hard to find out information on these fishes so that we can catch them or keep them.

Important food fishes include herrings and sardines (Clupeidae); anchovies (Engraulidae); salmons (Salmonidae); cods, Haddock, and Pollock (Gadidae); rockfishes (Scorpaenidae); Sauger, Walleye, and Yellow Perch (Percidae); drums and croakers (Sciaenidae); grunts (Haemulidae); snappers (Lutjanidae); Swordfish (Xiphiidae); mackerels and tunas (Scombridae); and many species of flatfishes such as Atlantic and Pacific halibuts (*Hippoglossus*), flounders, and soles in several different families (Pleuronectidae, Paralichthyidae, and Soleidae).

Ichthyologists acquire knowledge about fishes using a variety of techniques. Students in an ichthyology class in Nicaragua collect fishes with a seine net (top photo) and observe fish behavior using snorkeling gear in a small stream (bottom photo). Being out in nature is a major attraction of being an ichthyologist.

An increasingly important category of food fishes is those species raised in farming type operations (aquaculture) or in marine systems (mariculture). Important aquaculture species include Atlantic Salmon (*Salmo salar*, Salmonidae) that is being raised in many different parts of the world, and Milkfish (*Chanos chanos*, Chanidae) raised in brackish ponds in Asia. Freshwater species that are commonly farmed include carp, tilapia, Channel Catfish (*Ictalurus punctatus*, Ictaluridae), and Rainbow Trout (*Oncorhynchus mykiss*). Marine species include Cobia (*Rachycentron canadum*, Rachycentridae) and Yellowtail (*Seriola lalandi*, Carangidae). Recently, juvenile Atlantic Bluefin Tuna (*Thunnus thynnus*) in the Mediterranean Sea and Southern Bluefin Tuna (*Thunnus maccoyii*) in southern Australia are being caught, put into pens, and fattened for sale at very high prices to use in Japanese sushi restaurants.

Important sport fishes include Tarpon (*Megalops atlanticus*); Bonefish (*Albula vulpes*); trouts and salmons (Salmonidae); pike and Muskellunge (*Esox*, Esocidae); sunfishes and black basses (Centrarchidae); groupers and sea basses (Serranidae); Striped Bass (*Morone saxatilis*, Moronidae); dolphinfishes (*Coryphaena*, Coryphaenidae); billfishes (Istiophoridae); and Spanish mackerels and tunas (Scombridae).

Important pet fishes include many species of tetras (several families of Characiformes); loaches (superfamily Cobitoidea); catfishes (Siluriformes) such as the many species of armored catfishes (*Corydoras*, Callichthyidae) and the suckermouth catfishes (Loricariidae); minnows, carps, and barbs (Cyprinidae); killifishes (Cyprinodontiformes); cichlids (Cichlidae) such as the freshwater Oscar (*Astronotus*), angelfishes (*Pterophyllum*) and the Discus (*Symphysodon*); labyrinth fishes such as the Siamese Fighting Fish (*Betta splendens*) and the gouramis. Marine species kept in aquariums include damselfishes (Pomacentridae), lionfishes (*Pterois*, Scorpaenidae), wrasses (Labridae), anemone fishes (*Amphiprion*, *Premnas*; Pomacentridae), triggerfishes (Balistidae), and filefishes (Monacanthidae).

Which species of fishes are least well known?

The least known fishes are those with which we humans least interact, fishes that are not important for food, sport, or pets. This includes many small fishes such as minnows (Cyprinidae) and darters (Percidae) in our streams, gobies (Gobiidae) around coral reefs, and many deep-sea fishes that live below the depths that we can explore using scuba.

How do scientists tell fishes apart?

Ichthyologists find differences between individuals or populations of fishes. These differences may be in color patterns or shapes of the body or fins. Then ichthyologists may count fin rays, gill rakers, scales on a particular part of the body, or vertebrae or make measurements of the length of the fish compared with lengths of other body parts such as head length, length of the various fins, depth of the body, and so on. When one, or preferably more than one character, consistently differentiates one group of fishes from another, the ichthyologist may decide that these represent different species.

Internal characters are useful in sorting out species and determining their relationships. Ichthyologists study internal anatomy by dissection of fresh and preserved specimens; x-raying them, which produces radiographs; making skeletons of larger fishes; or by a process called clearing and staining, which makes the flesh transparent and the bones visible.

Ideas about relationships of fishes can now be tested using molecular characters—comparing mitochondrial or nuclear DNA to see if the molecular characters support the conclusions reached using morphological characters. Molecular data recently gave ichthyologists a clue as to the relationships of

Scuba diving with available equipment is safe only down to about 150 feet. Recent advances in diving technology, including computerized systems that adjust the gases breathed by a diver, have opened up deeper regions. The "twilight" zone on coral reefs below 150 feet houses many fish species previously unknown to science, such as this recently discovered Dr. Seuss Soapfish, *Belonoperca pylei*, a small (6 centimeter, 2.5 inch) member of the seabass family Serranidae. The fish has a bright yellow head and back, pink body, and orange spots. Photo courtesy of John E. Randall

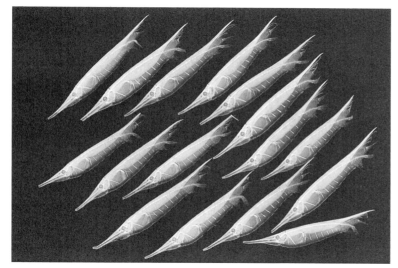

A radiograph (x-ray photograph) of a shrimpfish, *Centriscus scutatus*. Making x-rays allows an ichthyologist to view internal bones that show up as white lines in the photo. Each of the 16 x-rayed fish is about 10–12 centimeter (4–5 inches) long. Courtesy of Sandra Raredon

three deep-sea families of fishes with very different morphologies: Mirapinnidae (tapetails), Megalomycteridae (bignose fishes), and Cetomimidae (whalefishes). The nine genera and 20 species of what were traditionally called whalefishes and all 600-plus known individuals were captured below 1,000 meters (3,000 feet) and are females. Tapetails, as previously defined, comprise five species in three genera, and all 120 specimens are sexually immature and all but four collected in the upper 200 meters (600 feet). Bignose fishes comprise four monotypic (one species per genus) genera mostly collected below 1,000 meters, and all are males. Additional anatomical and ecological differences existed among the presumed families. Larvae (tapetails) have small upturned mouths and feed mostly on small crustaceans. Females (whalefishes) have enormous mouths with long jaws. Males (bignoses) did not feed, having lost their stomach and esophagus as they matured, and apparently lived off a massive liver.

That each "family" contained only females, larvae, or males and differed in other respects caused several researchers to suspect that the traditional classifications were in error. Because larvae in many deep-sea species differ greatly from adults of the same species, as do males and females (see chapter 6, "How do fishes reproduce?" and "What is a baby fish called?"), there was reason to believe that a similar phenomenon existed here. Fortunately, a recent deep-sea collecting expedition captured animals that were apparently transforming between different stages. Clearing and staining of these and other specimens showed dramatic changes in the skeleton with growth, confirming the suspicion of mistaken identities. These changes convinced a team of seven ichthyologists from three continents, led by David Johnson and John Paxton, to unite the three fish families into one family, the whalefishes (Cetomimidae).

How do you become an ichthyologist?

The science of ichthyology, as with other areas of biology that focus on particular groups of organisms, requires a broad background in a variety of related fields. Anyone wishing to pursue a career in ichthyology needs to take college courses not only in biology but also in math, chemistry, amd physics, to name a few (including English composition). The most successful researchers are people who can draw upon the insight gained from a broad background to answer meaningful research questions. After graduation from a university or college, a student can then specialize in a graduate program designed to conduct research in the lab or field focused on systematics, ecology, biogeography, physiology, behavior, or conservation.

But to study fishes and contribute to our knowledge of them does not require a science degree. Many "cititzen scientists" whose daily lives are unrelated to ichthyology contribute substantially to the study and conservation of fishes. Opportunities abound to collect, observe, and conserve fishes and their habitats. A good place to start looking for resources are the websites listed in appendix B (Some Organizations That Promote Ichthyology and the Conservation of Fishes).

Appendix A

The Classification of Fishes

The following is based on Nelson (2006), with some modifications from Wiley and Johnson (2010). An online, updated Catalog of Fishes is maintained by the California Academy of Sciences at: http://researcharchive.calacademy.org/research/Ichthyology/catalog/fishcatmain.asp. Species mentioned in this book are listed under their respective family.

Grade Teleostomi (Osteichthyes). Bony fishes
Class Sarcopterygii. Lobe-finned fishes (and tetrapods)
  Order Coelacanthiformes. Coelacanths
    Latimeriidae. 2 species of coelacanths
      Latimeria chalumnae. African Coelacanth
      Latimeria manadoensis. Indonesian Coelacanth
  Order Ceratodontiformes. Living lungfishes
    Ceratodontidae. Australian Lungfish
      Neoceratodus forsteri. Australian Lungfish
    Lepidosirenidae. South American Lungfish
      Lepidosiren paradoxa. South American Lungfish
    Protopteridae. 4 species of African lungfishes
      Protopterus annectens. African Lungfish

Class Actinopterygii. Ray-finned fishes
  Order Polypteriformes.
    Polypteridae. Bichirs
      Polypterus. 15 species of African bichirs
      Erpetoichthys calabaricus. African Reedfish
  Order Acipenseriformes. Sturgeons and paddlefishes
    Acipenseridae. 25 species of sturgeons
      Acipenser brevirostris. Shortnose Sturgeon
      Acipenser fulvescens. Lake Sturgeon
      Acipenser sturio. Baltic Sturgeon
      Huso huso. Beluga Sturgeon
    Polyodontidae. Paddlefishes
      Polyodon spathula. North American Paddlefish
      Psephurus gladius. Chinese Paddlefish

  Order Lepisosteiformes. Gars
    Lepisosteidae. 7 species of gars
      Atractosteus spatula. Alligator Gar
      Lepisosteus. Gars
  Order Amiiformes.
    Amiidae.
      Amia calva. Bowfin
  Order Hiodontiformes
    Hiodontidae. 2 species of mooneyes
  Order Osteoglossiformes. Bonytongues
    Osteoglossidae. 8 species of bonytongues
      Arapaima gigas. South American Pirarucu
      Osteoglossum bicirrhosum. South American Arawana
      Pantodon buchholzi. African Freshwater Butterflyfish
      Scleropages formosus. Golden Dragonfish
    Notopteridae. 8 species of featherfin knifefishes
    Mormyridae. About 200 species of African elephantfishes
      Gnathonemus petersi. Peters' Elephantfish
    Gymnarchidae.
      Gymnarchus niloticus. African Aba
  Order Elopiformes. 8 species
    Elopidae. Tenpounders and Ladyfish
    Megalopidae. Tarpons
      Megalops atlanticus. Atlantic Tarpon
  Order Albuliformes. 30 species of bonefishes
    Albulidae. 5 species of bonefishes
      Albula vulpes. Atlantic Bonefish
    Halosauridae. 15 species of halosaurs
    Notacanthidae. 10 species of spiny eels

Order Anguilliformes. 791 species of eels
    Anguillidae. 15 species of freshwater eels
        Anguilla anguilla. European Freshwater Eel
        Anguilla rostrata. North American Freshwater Eel
    Heterenchelyidae. 8 species of mud eels
    Moringuidae. 6 species of spaghetti eels
    Chlopsidae. 18 species of false moray eels
    Myrocongridae. 4 species of myroconger eels
    Muraenidae. About 185 species of moray eels
        Gymnothorax funebris. Green Moray
    Synaphobranchidae. About 32 species of cutthroat eels
    Ophichthidae. About 290 species of snake eels
    Colocongridae. 5 species of shorttail eels
    Derichthyidae. 3 species of longneck eels
    Muraenosocidae. About 8 species of pike congers
    Nemichthyidae. About 18 species of snipe eels
    Congridae. About 160 species of conger eels
    Nettastomidae. About 38 species of duckbill eels
    Serrivomeridae. About 20 species of sawtooth eels
Order Saccopharyngiformes. 28 species
    Cyematidae. 2 species of bobtail snipe eels
    Saccopharyngidae. About 10 species of swallowers
    Eurypharyngidae. 1 species of Gulper or Pelican Eel
    Monognathidae. About 15 species of onejaw gulpers
Order Clupeiformes. About 364 species of herring-like fishes
    Denticipitidae. Denticle Herring
    Pristigasteridae. 34 species of longfin herrings
    Engraulidae. 139 species of anchovies
    Chirocentridae. 2 species of wolf herrings
    Clupeidae. About 188 species of herrings
        Alosa pseudoharengus. Alewife.
        Alosa sapidissima. American Shad
        Brevoortia tyrannus. Atlantic Menhaden
        Clupea harengus. Atlantic Herring

Order Gonorhynchiformes. 37 species
    Chanidae. Milkfish
        Chanos chanos. Milkfish
    Gonorhynchidae. 5 species of beaked sandfishes
    Kneriidae. About 30 species of knerias
    Phractolaemidae. African Snake Mudhead
Order Cypriniformes. About 3,268 species
    Cyprinidae. About 2,420 species of minnows and carps
        Barbus. Barbs
        Cyprinella. Shiners
        Carassius auratus. Goldfish
        Cyprinus carpio. Carp
        Danio rerio. Zebrafish
        Erimonax monachus. Spotfin Chub
        Exoglossum maxillingua. Cutlip Minnow
        Garra rufa
        Labeo
        Nocomis. 7 species of chubs
        Nocomis leptocephalus. Bluehead Chub
        Nocomis micropogon. River Chub
        Paedocypris progenetica. Tiny Indonesian Minnow
        Phoxinus phoxinus. European Minnow
        Pimephales promelas. Fathead Minnow
        Probarbus jullieni. Julien's Golden Carp
        Ptychocheilus lucius. Colorado Pikeminnow
        Puntius. Asian barbs
        Rhodeus. Bitterlings
        Tor putitora. Indian Mahseer
        Semotilus. 4 species of chubs
    Psilorhynchidae. 6 species of mountain carps
    Gyrinocheilidae. 3 species of algae eaters
    Catostomidae. 72 species of suckers
        Chasmistes cujus. Cui-ui Sucker
        Ictiobus cyprinellus. Bigmouth Buffalo
        Moxostoma. 22 species of redhorse suckers
        Moxostoma robustum. Robust Redhorse
    Cobitidae. About 177 species of loaches
    Balitoridae. More than 590 species of river loaches
        Cryptotora thamicola. Torrentfish
Order Characiformes. At least 1,674 species of characins
    Distichodontidae. About 90 species of distichodontids

                                           Classification of Fishes

Citharinidae. 8 species of African citharinids
Parodontidae. About 21 described species of parodontids
Curimatidae. About 95 species of toothless characiforms
Prochilodontidae. About 21 species of flannel-mouths
Anostomidae. At least 137 species of toothed headstanders
Chilodontidae. 7 species of headstanders
Crenuchidae. 74 species of South American darters
Hemiodontidae. About 28 described species of hemiodontids
Alestiidae. About 110 species of African tetras
Gasteropelecidae. 9 species of freshwater hatchetfishes
Characidae. More than 962 species of characins
　Astyanax jordani. Mexican Blind Cavefish
　Astyanax mexicanus. Mexican Tetra
　Brycon. 41 species of characins
　Colossoma
　Myleus. 15 species of plant-eating characins
　Paracheirodon axelrodi. Cardinal Tetra
　Pygocentrus. 4 species of red-bellied piranhas
　Serrasalmus. 28 species of piranhas
Acestrorhynchidae. 15 species of acestrorhynchids
Cynodontidae. 14 species of cynodontids
Erythrinidae. About 14 species of trahiras
Lebiasinidae. 61 species of pencil fishes
　Copella arnoldi. Spraying Characin
Ctenoluciidae. 7 species of pike-characins
Hepsetidae. African Pike
Order Siluriformes. 36 families and at least 3,093 species of catfishes
Diplomystidae. 6 species of velvet catfishes
Cetopsidae. 41 species of whalelike catfishes
Amphiliidae. 66 species of loach catfishes
Trichomycteridae. About 207 species of parasitic catfishes
　Vandellia. 3 species of candirus
　Vandellia cirrhosa. Candiru
Nematogenyidae. 2 species of mountain catfishes

Callichthyidae. 194 species of armored catfishes
　Corydoras. 153 species
Scoloplacidae. 4 species of spiny dwarf catfishes
Astroblepidae. 54 species of climbing catfishes
Loricariidae. About 716 species of suckermouth catfishes
　Panaque. 3 species of "plecostomus"
　Pterygoplichthys. 14 species of "plecos" or "plecostomus"
Amblycipitidae. About 27 species of torrent catfishes
Akysidae. 47 species of stream catfishes
Sisoridae. At least 112 species of sisorid catfishes
Aspredinidae. 36 species of banjo catfishes
Pseudopimelodidae. 29 species of bumblebee catfishes
Heptapteridae. 189 species of heptapterid catfishes
Cranoglanididae. 3 species of armorhead catfishes
Ictaluridae. About 64 species of North American catfishes
　Ameiurus. 7 species of bullheads
　Ictalurus punctatus. Channel Catfish
　Noturus. 28 species of madtom catfishes
Mochokidae. 188 species of upside-down catfishes
　Synodontis multipunctata. Cuckoo Catfish
Anchariidae. 5 species of anchariid catfishes
Doradidae. About 77 species of thorny catfishes
Auchenipteridae. About 94 species of driftwood catfishes
Siluridae. About 94 species of sheatfishes
Malapteruridae. 19 species of electric catfishes
　Malapterurus. 16 species of electric catfishes
Auchenoglanididae. 28 species of auchenoglanidid catfishes
Chacidae. 3 species of squarehead catfishes
Plotosidae. About 35 species of eeltail catfishes

Plotosus lineatus. Venomous Catfish
Tandanus. Australian eel-tailed catfishes
Clariidae. About 113 species of airbreathing catfishes
Clarias batrachus. Walking Catfish
Heteropneustidae. About 3 species of airsac catfishes
Austroglanidae. 3 species of austroglanid catfishes
Claroteidae. About 83 species of claroteid catfishes
Ariidae. About 133 species of sea catfishes
Schilbeidae. 62 species of schilbeid catfishes
Pangasiidae. About 30 species of shark catfishes
Pangasianodon gigas. Mekong Giant Catfish
Bagridae. About 189 species of bagrid catfishes
Pimelodidae. At least 93 species of long-whiskered catfishes
Lacantuniidae. Monotypic, Chiapas Catfish
Order Gymnotiformes. About 134 species of American knifefishes
Gymnotidae. 33 species of nakedback knifefishes
Electrophorus electricus. Electric Eel
Rhamphichthyidae. 12 species of sand knifefishes
Hypopomidae. 16 species of bluntnose knifefishes
Sternopygidae. About 28 species of glass knifefishes
Apteronotidae. About 45 species of ghost knifefishes
Order Argentiniformes. About 202 species of marine smelts
Argentinidae. About 23 species of argentines
Opisthoproctidae. About 11 species of barreleyes
Microstomatidae. About 38 species of pencilsmelts
Platytroctidae. 37 species of tubeshoulders
Bathylaconidae. 3 species of bathylaconids
Alepocephalidae. At least 90 species of slickheads
Order Osmeriformes. 88 species of smelts
Osmeridae. 31 species of smelts
Plecoglossus altivelis. Ayu

Retropinnidae. About 5 species of New Zealand smelts
Galaxiidae. 52 species of galaxiids
Lepidogalaxias salamandroides. Salamanderfish
Order Salmoniformes. About 66 species
Salmonidae. Trouts, salmons, and whitefishes
Oncorhynchus gorbuscha. Pink Salmon
Oncorhynchus mykiss. Rainbow Trout
Oncorhynchus mykiss aguabonita. Golden Trout
Oncorhynchus nerka. Sockeye Salmon
Oncorhynchus tshawytscha. Chinook Salmon
Salmo salar. Atlantic Salmon
Salmo trutta. Brown Trout
Salvelinus namaycush. Lake Trout
Order Esociformes. 10 species of pickerels
Esocidae. 5 species of pikes and pickerels
Esox lucius. Northern Pike
Umbridae. 5 species of mudminnows
Umbra pygmaea. Eastern Mudminnow
Order Stomiiformes. About 391 species of dragonfishes
Diplophidae. 8 species
Gonostomatidae. 23 species of bristlemouths
Sternoptychidae. About 67 species of marine hatchetfishes
Phosichthyidae. About 20 species of lightfishes
Stomiidae. About 273 species of barbeled dragonfishes
Malacosteus. Genus of loosejaws
Order Atelopodiformes. 1 family
Atelopodidae. About 12 species of jellynose fishes
Order Aulopiformes. About 236 species of lizardfishes
Paraulopidae. 10 species of cucumber fishes
Aulopidae. About 10 species of flagfins
Pseudotrichonotidae. Sanddiving Lizardfish
Synodontidae. About 57 species of lizardfishes
Bathysauroididae. Bathysauroid
Chloropthalmidae. About 19 species of greeneyes

Bathysauropsidae. 3 species of
    bathysauropsids
Notosudidae. 19 species of waryfishes
Ipnopidae. 29 species of deepsea tripod
    fishes
Scopelarchidae. 17 species of pearleyes
Evermannellidae. 7 species of sabertooth
    fishes
Alepisauridae. 3 species of lancetfishes
Paralepididae. About 56 species of
    barracudinas
Bathysauridae. 2 species of deepsea
    lizardfishes
Giganturidae. 2 species of telescopefishes
Order Myctophiformes. About 246 species of
    lanternfishes
Neoscopelidae. 6 species of blackchins
Myctophidae. At least 240 species of
    lanternfishes
    Diaphus. Genus of headlight fishes
    Myctophum. Genus of lanternfishes
Order Lampriformes. 21 species
Veliferidae. 2 monotypic genera of velifers
Lampridae. 2 species of opahs
Stylephoridae. Probably 1 species of tube-
    eye
Lophotidae. 3 species of crestfishes
Radiicephalidae. Monotypic, tapertail.
Trachipteridae. About 10 species of
    ribbonfishes
Regalecidae. 2 monotypic genera of
    oarfishes
    Regalecus glesne. Oarfish
Order Polymixiiformes. 1 genus of beardfishes
Polymixiidae. 10 species of beardfishes
Order Percopsiformes. 9 species
Percopsidae. 2 species of North American
    trout-perches
Aphredoderidae. North American Pirate
    Perch.
Amblyopsidae. 6 species of North American
    cave fishes
Order Gadiformes. About 555 species of cods
Muraenolepidae. 4 species of eel cods
Bregmacerotidae. At least 15 species of
    codlets
Euclichthyidae. Monotypic, Eucla Cod
Macrouridae. About 350 species of
    grenadiers or rattails

Moridae. About 105 species of deep-sea cods
Melanonidae. 2 species of pelagic cods
Merlucciidae. 22 species of hakes
Phycidae. 25 species of phycid hakes
Gadidae. About 31 species of cods
    Gadus morhua. Atlantic Cod
    Pollachius virens. Pollock
Order Zeiformes. 32 species of dories
Cyttidae. 3 species of lookdown dories
Oreosomatidae. 9 or 10 species of oreos
Parazenidae. 3 species of smooth dories
Zeniontidae. About 7 species of armoreye
    dories
Grammicolepidae. 2 monotypic genera of
    tinselfishes
Zeidae. 5 species of dories
    Zeus faber. John Dory
Order Stephanoberyciformes. More than 75
    species of pricklefishes
Melamphaeidae. 36 species of bigscale fishes
Stephanoberycidae. 3 monotypic genera of
    pricklefishes
Hispidoberycidae. 1 species
Gibberichthyidae. 2 species of gibberfishes
    Gibberichthys pumilus. Gibberfish
Rondeletiidae. 2 species of redmouth
    whalefishes
Barbourisidae. Red whalefish
Cetomimidae (including Mirapinnidae and
    Megalomycteridae)
    about 20 species of whalefishes, tapetails,
    and bignoses
Order Beryciformes. 144 species of alfonso
    squirrelfishes
Anoplogasteridae. 2 species of fangtooths
Diretmidae. 4 species of spinyfins
Anomalopidae. 8 species of flashlight fishes
Monocentridae. 4 species of pinecone
    fishes
Trachichthyidae. About 39 species of
    roughies
    Hoplostethus atlanticus. Orange Roughy
Berycidae. About 9 species of alfonsinos
Holocentridae. About 78 species of
    squirrelfishes
Percomorpha
Order Mugiliformes. 1 family and about 17
    genera
Mugilidae. 72 species of mullets

Order Atheriniformes. About 312 species of
  silversides
  Atherinopsidae. About 108 species of New
    World silversides
    Leuresthes tenuis. California Grunion
    Menidia. 29 species of silversides
  Notocheiridae. 6 species of surf sardines
  Melanotaeniidae. 113 species of
    rainbowfishes
    Pseudomugil. 13 species of blue eyes
  Atherionidae. 3 species of pricklenose
    silversides
  Phallostethidae. 22 species of priapiumfishes
  Atherinidae. About 60 species of Old World
    silversides
Order Beloniformes. At least 227 species
  Adrianichthyidae. 28 species of
    adrianichthyids
    Oryzias latipes. Medaka
  Belonidae. 34 species of needlefishes
  Scomberesocidae. 4 species of sauries
  Zenarchopteridae. 54 species of halfbeaks
    Dermogenys. 13 species
    Hemirhamphodon. 6 species
    Nomorhamphus. 16 species
  Hemiramphidae. About 55 species of
    halfbeaks
  Exocoetidae. About 50 species of flyingfishes
Order Cyprinodontiformes. About 1,013 species
  of killifishes
  Aplocheilidae. At least 7 species of Asian
    rivulines
  Nothobranchidae. About 250 species of
    African rivulines
    Nothobranchius. Annual fish
  Rivulidae. 236 described species of New
    World rivulines
    Kryptolebias marmoratus. (formerly
      Rivulus)
  Profundulidae. 5 species of Middle
    American killifishes
  Goodeidae. About 40 species of goodeids
    and splitfins
    Ameca splendens. Butterfly Splitfin
  Fundulidae. About 50 species of topminnows
  Valenciidae. 2 species of Valencia toothcarps
  Cyprinodontidae. 104 species of pupfishes
    Cyprinodon diabolis. Devils Hole
      Pupfish

Anablepidae. 15 species of four-eyed fishes
    Anableps. 3 species of four-eyed fishes
  Poeciliidae. About 304 species of livebearers
    Poecilia reticulata. Guppy
    Poeciliopsis. 21 species
    Xiphophorus. 25 species of swordtails and
      platys
Order Elassomatiformes. 1 family
  Elassomatidae. 6 species of pygmy sunfishes
Order Gasterosteiformes. 11 families and 278
  species
  Hypoptychidae. Monotypic, Sand Eel.
  Aulorhynchidae. 2 species of tubesnouts
    Gasterosteidae. 8 species of sticklebacks
  Indostomidae. 3 species of armored
    sticklebacks
  Pegasidae. 5 species of seamoths
  Solenostomidae. 4 or 5 species of ghost
    pipefishes
  Syngnathidae. About 232 species of
    pipefishes and seahorses
    Hippocampus bargibanti. Pygmy Seahorse
    Phycodurus eques. Leafy Seadragon
  Aulostomidae. 1 genus of trumpetfishes
    Aulostomus. 3 species of trumpetfishes
  Fistulariidae. 1 genus of cornetfishes
    Fistularia. 4 species of cornetfishes
  Macroramphosidae. About 11 species of
    snipefishes
  Centriscidae. About 4 species of
    shrimpfishes
    Centriscus scutatus. Shrimpfish
Order Synbranchiformes. 3 families and about
  99 species
  Synbranchidae. 17 species of swamp eels
  Chaudhuriidae. 9 species of earthworm
    eels
  Mastacembelidae. About 73 species of spiny
    eels
Order Dactylopteriformes. 1 family
  Dactylopteridae. 7 species of flying gurnards
Order Scorpaeniformes. 11 families and about
  600 species
  Scorpaenidae. About 418 species of
    scorpionfishes
    Pterois. Lionfishes
    Scorpaena aleutianus. Rougheye Rockfish
    Scorpaena mystes. Pacific Spotted
      Scorpionfish

Classification of Fishes

Sebastes pinniger. Canary Rockfish
Synanceia. 5 species of stonefishes
Caracanthidae. About 4 species of orbicular velvetfishes
Aploactinidae. About 38 species of velvetfishes
Pataecidae. 3 monotypic genera of Australian prowfishes
Gnathanacanthidae. Red Velvetfish
Congiopodidae. 15 species of pigfishes
Triglidae. About 105 species of searobins and gurnards
Peristediidae. About 36 species of armored searobins
Bembridae. About 10 species of deepwater flatheads
Platycephalidae. About 65 species of flatheads
Hoplichthyidae. About 10 species of ghost flatheads
Order Perciformes. 160 families with 10,000+ species
Suborder Percoidei. 71 families with about 2, 806 species
Centropomidae. 12 species of snooks
Latidae. 9 species
   Lates calcarifer. Barramundi
   Lates japonicus. Japanese Snook
Ambassidae. About 46 species of Asiatic glassfishes
Moronidae. 8 species of temperate basses
   Morone saxatilis. Striped Bass
Percichthyidae. About 34 species of temperate perches
Perciliidae. 2 species of southern basses
Acropomatidae. About 31 species of lanternbellies
Symphysanodontidae. At least 6 species of slopefishes
Polyprionidae. About 5 species of wreckfishes
Serranidae. About 475 species of sea basses
   Belonoperca pylei. Dr. Seuss Fish
   Epinephelus guttatus. Red Hind
   Epinephelus lanceolatus. Giant Grouper
   Epinephelus striatus. Nassau Grouper
   Hypoplectrus. Hamlets
   Paralabrax clathratus. Kelp Bass
   Rypticus. Soapfishes

Centrogeniidae. Monotypic, False Scorpionfish
Ostracoberycidae. 3 species of ostracoberycids
Callanthiidae. About 12 species of groppos
Pseudochromidae. At least 119 species of dottybacks
Grammatidae. 12 species of basslets
Plesiopidae. About 46 species of roundheads
Notograptidae. Perhaps 3 species of bearded eelblennies
Opistognathidae. More than 78 species of jawfishes
Dinopercidae. 2 species of cavebasses
Banjosidae. Monotypic, Banjofish
Centrarchidae. 31 species of sunfishes and black basses
   Ambloplites. Rock basses
   Lepomis cyanellus. Green Sunfish
   Lepomis macrochirus. Bluegill Sunfish
   Micropterus dolomieu. Smallmouth Bass
   Micropterus salmoides. Largemouth Bass
   Pomoxis nigromaculatus. Black Crappie
Percidae. 201 species of darters and perches
   Etheostoma. 136 species of darters
   Perca flavescens. Yellow Perch
   Percina aurantiaca. Tangerine Darter
   Percina tanasi. Snail Darter
   Sander. 5 species of pikeperches
Priacanthidae. About 18 species of bigeyes
   Apogonidae. Roughly 273 species of cardinalfishes
   Pterapogon kauderni. Banggai Cardinalfish
Kurtidae. 2 species of nurseryfishes
Epigonidae. Roughly 25 species of deepwater cardinalfishes
Sillaginidae. About 31 species of sillagos or whitings
Malacanthidae. About 40 species of tilefishes
   Malacanthus. 3 species of sand tilefishes
Lactariidae. Monotypic, False Trevally
Dinolestidae. Monotypic. Long-finned Pike
   Dinolestes lewini. Australian Long-finned Pike
Scombropidae. About 3 species of gnomefishes
Pomatomidae. Monotypic
   Pomatomus saltatrix. Bluefish

Menidae. Monotypic, moonfish
   Mene maculata. Moonfish
Leiognathidae. About 30 species of
   ponyfishes
   Leiognathus. Genus of ponyfishes
Bramidae. 22 species of pomfrets
Caristiidae. About 5 species of manefishes
Emmelichthyidae. 15 species of rovers
Lutjanidae. About 105 species of snappers
Caesionidae. 20 species of fusiliers
Lobotidae. About 5 species of tripletails
Gerreidae. About 44 species of mojarras
Haemulidae. About 145 species of grunts
   Haemulon flavolineatum. French Grunt
Inermiidae. 2 species of bonnetmouths
Polynemidae. 41 species of threadfins
Sciaenidae. About 270 species of croakers
   and drums
   Aplodinotus grunniens. Freshwater Drum
   Bairdiella. Genus of croakers
   Cynoscion. Genus of seatrouts
   Totoaba macdonaldi. Giant Totoaba
Mullidae. About 62 species of goatfishes
   Mulloidichthys martinicus. Yellow Goatfish
Pempheridae. About 26 species of sweepers
Glaucosomatidae. 4 species of pearl perches
Leptobramidae. Monotypic, Beachsalmon
Bathyclupeidae. About 5 species of
   bathyclupeids
Monodactylidae. About 5 species of
   moonfishes
Toxotidae. 1 genus and 6 species of
   archerfishes
   Toxotes jaculator. Archerfish
Arripidae. 1 genus and 4 species of
   Australasian salmons
Dichistidae. 1 genus and 2 species of galjoen
   fishes
Kyphosidae. 45 species of sea chubs
Drepaneidae. 1 genus and 2 or 3 species of
   sicklefishes
Chaetodontidae. About 122 species of
   butterflyfishes
   Chaetodon lunula. Raccoon Butterflyfish
Pomacanthidae. About 82 species of
   angelfishes
   Pomacanthus imperator. Emperor
   Angelfish
Enoplosidae. Monotypic, Oldwife

Pentacerotidae. About 12 species of
   armorheads
Nandidae. 21 species of Asian leaffishes
Polycentridae. 4 species of Afro-American
   leaffishes
Terapontidae. About 48 species of grunters
Kuhliidae. 1 genus with 10 species of
   flagtails
Oplegnathidae. 1 genus with about 7 species
   of knifejaws
Cirrhitidae. About 33 species of hawkfishes
Chironemidae. About 5 species of kelpfishes
Cheilodactylidae. About 22 species of
   morwongs
Aplodactylidae. About 5 species of
   marblefishes
Cepolidae. 19 species of bandfishes
Sphyraenidae. 1 genus and about 21 species
   of barracudas
   Sphyraena barracuda. Great Barracuda
   Sphyraena genie. Chevron Barracuda
Suborder Sparoidei. 4 families, 226 species
   Nemipteridae. About 64 species of threadfin
      breams
   Lethrinidae. About 39 species of emperors
   Sparidae. About 115 species of porgies
      Archosargus probatocephalus. Sheepshead.
   Centracanthidae. 8 species of picarel porgies
Suborder Carangoidei. 5 families, 152 species
   Nematistiidae. Monotypic, Roosterfish
   Coryphaenidae. 2 species of dolphinfishes
      Coryphaena hippurus. Common
         Dolphinfish or Mahi-mahi
   Rachycentridae. Monotypic
      Rachycentron canadum. Cobia
   Echeneidae. 8 species of remoras or
      sharksuckers
   Carangidae. About 140 species of jacks and
      pompanos
      Oligoplites. Leatherjackets
      Seriola lalandi. Yellowtail
Suborder Labroidei. 6 families with 2,274
      species
   Cichlidae. At least 1,300 species of cichlids
      Astronotus ocellatus. Oscar
      Cichla ocellaris. Peacock Bass
      Dimidiochromis. Lake Malawi Eyebiter
      Docimodus johnstoni. Lake Malawai
         Finbiter

Classification of Fishes

Julidochromis

Lamprologus

Neolamprologus

Nimbochromis livingstonii. Kalingo

Pterophyllum scalare. Freshwater Angelfish

Rhamphochromis

Sarotherodon galilaeus. Tilapia

Symphysodon. Discus

Tyrannochromis macrostoma. Lake Malawi predator

Embiotocidae. About 23 species of viviparous surfperches

Pomacentridae. About 348 species of damselfishes

Amphiprion (including *Premnas*). Anenomefishes

Chromis punctipinnis. Blacksmith

Dascyllus. Damselfishes

Hypsypops rubicundus. Garibaldi

Pomacentrus. Damselfishes

Stegastes leucostictus. Beaugregory Damselfish

Labridae. Wrasses (including parrotfishes, formerly Scaridae)

Cheilinus undulatus. Humphead Wrasse

Coris. Coris wrasses

Epibulus insidator. Sling-jaw Wrasse

Labroides dimidiatus. A species of cleaner wrasse

Oxyjulis californica. Senorita

Scarus croicensis. Striped Parrotfish

Scarus guacamaia. Rainbow Parrotfish

Symphodus melops. European Corkwing Wrasse

Thalassoma bifasciatum. Bluehead Wrasse

Odacidae. 12 species of cales

Suborder Notothenioidei. 8 families with 125 species

Bovichtidae. About 11 species of temperate icefishes

Pseudophritidae. Monotypic, Catadromous Icefish

Eliginopidae. Monotypic, Patagonian Blenny

Nototheniidae. About 50 species of cod icefishes

Dissostichus eleginoides. Patagonian Toothfish or Chilean Seabass

Harpagiferidae. About 6 species of spiny plunderfishes

Artedraconidae. About 25 species of barbeled plunderfishes

Bathydraconidae. 16 species of Antarctic dragonfishes

Channichthyidae. 15 species of crocodile icefishes

Suborder Trachinoidei. 12 families with 237 species

Chiasmodontidae. About 15 species of swallowers

Champsodontidae. About 13 species of gapers

Trichodontidae. 2 species of sandfishes

Pinguipedidae. About 54 species of sandperches

Cheimarrhichthyidae. Monotypic, New Zealand Torrentfish

Trichonotidae. About 8 species of sanddivers

Creediidae. About 16 species of sand burrowers

Percophidae. About 44 species of duckbills

Leptoscopidae. 5 species of southern sandfishes

Ammodytidae. 23 species of sand lances

Trachinidae. 6 species of weeverfishes

Uranoscopidae. About 30 species of stargazers

Astroscopus. Genus of electric stargazers

Suborder Pholidichthyoidei. 1 family

Pholidichthyidae. 2 species of convict blennies

Suborder Blennioidei. 6 families with at least 818 species

Tripterygiidae. About 150 species of triplefin blennies

Notoclinops. Genus of triplefin blennies

Notoclinops segmentatus. Blue-eyed Triplefin

Dactyloscopidae. 43 species of sand stargazers

Blenniidae. About 360 species of combtooth blennies

Aspidontus. Genus of saber-toothed blennies

Plagiotremus. Genus of saber-toothed blennies

Clinidae. 74 species of kelp blennies

Labrisomidae. About 105 species of
labrisomid blennies

Chaenopsidae. About 86 species of tube
blennies

Chaenopsis ocellata. Bluethroat Pikeblenny

Suborder Gobiodei. 9 families, more than
2,211 species

Rhyacicthyidae. 2 species of loach gobies

Odontobutidae. About 15 species of
freshwater sleepers

Eleotridae. About 155 species of sleepers

Xenisthmidae. About 12 species of
xenisthmids

Kraemeriidae. About 8 species of sand
gobies

Gobiidae. At least 1,950 species of gobies

Brachygobius. Bumblebee gobies

Eviota sigillata. Short-lived pygmy coral
reef goby

Gobiodon. Clown gobies

Gobiosoma. Genus of cleaner gobies

Mistichthys. Genus of tiny gobies

Pandaka. Genus of tiny gobies

Periopthalmus. Genus of mudskippers

Trimmaton nanus. Indian Ocean goby

Microdesmidae. About 30 species of
wormfishes

Ptereleotridae. About 36 species of
dartfishes

Schindleriidae. Infantfishes

Schindleria. 3 species of infantfishes

Suborder Acanthuroidei. 6 families and about
129 species

Ephippidae. About 16 species of spadefishes

Scatophagidae. 4 species of scats

Scatophagus. Genus of scats

Siganidae. 27 species of rabbitfishes

Luvaridae. Monotypic, Louvar

Zanclidae. Monotypic, Moorish Idol

Acanthuridae. About 80 species of
surgeonfishes

Naso. About 16 species of unicornfishes

Suborder Xiphioidei. 10 species

Xiphiidae. Monotypic, Swordfish

Xiphias gladius. Swordfish

Istiophoridae. Billfishes and marlins

Istiompax indica. Black Marlin

Istiophorus platypterus. Sailfish

Makaira nigricans. Blue Marlin

Suborder Scombroidei. 114 species

Scombrolabracidae. Monotypic, Longfin
Escolar

Gempylidae. About 24 species of snake
mackerels

Trichiuridae. About 39 species of
cutlassfishes

Scombridae. 51 species of mackerels and
tunas

Gymnosarda unicolor. Dogtooth Tuna

Scomber. 4 species of mackerels

Scomberomorus regalis. Cero Mackerel

Thunnus albacares. Yellowfin Tuna

Thunnus maccoyii. Southern Bluefin Tuna

Thunnus thynnus. Atlantic Bluefin Tuna

Suborder Stromateoidei. 6 families and about
70 species

Amarsipidae. Monotypic, Amarsipa

Centrolophide. About 28 species of
medusafishes

Nomeidae. About 16 species of driftfishes

Nomeus gronovii. Man-of-War Fish

Ariommatidae. About 7 species of
ariommatids

Tetragonuridae. 3 species of squaretails

Stromateidae. About 15 species of
butterfishes

Suborder Icosteoidei

Icosteidae. monotypic, Ragfish

Suborder Caproidei. 1 family

Caproidae. About 11 species of boarfishes

Order Gobiesociformes. 3 families and about 36
genera

Suborder Gobiesocoidei. 1 family

Gobiesocidae. About 140 species of
clingfishes

Suborder Callionymoidei. 2 families

Callionymidae. About 182 species of
dragonets

Draconettidae. About 12 species of slope
dragonets

Order Anabantiformes. 3 families and about 120
species of labyrinth fishes

Suborder Anabantoidei

Anabantidae. About 33 species of climbing
gouramis

Betta splendens. Siamese Fighting Fish

Helostomatidae. Monotypic, Kissing
Gourami

Osphronemidae. About 86 species of gouramis
Suborder Channoidei. 1 family
Channidae. 29 species of Snakeheads
Channa micropeltes. Giant Snakehead
Order Cottiformes, 2 suborders
Suborder Cottoidei, 14 families, about 670 species
Anoplopomatidae. 2 monotypic genera of sablefishes
Anoplopoma fimbria. Sablefish
Hexagrammidae. About 12 species of greenlings
Normanichthyidae. Monotypic
Rhamphocottidae. Monotypic, Grunt Sculpin
Ereuniidae. 3 species of deepwater sculpins
Cottidae. About 275 species of sculpins
Hemilepidotus. Red Irish Lord
Comephoridae. 2 species of Lake Baikal oilfishes
Abysocottidae. About 22 species of deepwater Baikal sculpins
Hemitripteridae. 8 species of searavens
Agonidae. 47 species of poachers
Psychrolutidae. About 35 species of fathead sculpins
Bathylutichthyidae. Monotypic, Antarctic Sculpin
Cyclopteridae. 28 species of lumpfishes
Liparidae. About 334 species of snailfishes
Careproctus. Genus of snailfishes
Liparis. Genus of snailfishes
Suborder Zoarcoidei. 9 families, 340 species
Bathymasteridae. 7 species of ronquils
Zoarcidae. About 230 species of eelpouts
Stichaeidae. About 76 species of pricklebacks
Cryptacanthodidae. 1 genus, 4 species of wrymouths
Pholidae. About 15 species of gunnels
Anarhichadidae. 5 species of wolffishes
Anarrhichthys ocellatus. Wolf-eel
Ptilichthyidae. Monotypic, Quillfish
Zaproidae. Monotypic, Prowfish
Scytalinidae. Monotypic, Graveldiver
Order Ophidiiformes. About 385 species of cusk-eels
Carapidae. 31 species of pearlfishes

Encheliophis. Genus of commensal pearlfishes
Ophidiidae. About 222 species of cusk-eels
Abyssobrotula galatheae. Deepsea Cusk-eel
Bythitidae. About 107 species of viviparous brotulas
Aphyonidae. 22 species of aphyonids
Parabrotulidae. 3 species of false brotulas
Order Batrachoidiformes. 22 genera of toadfishes
Batrachoididae. 78 species of toadfishes
Opsanus tau. Oyster Toadfish
Porichthys. 14 species of midshipmen
Order Lophiiformes. 313 species of anglerfishes
Lophiidae. 25 species of goosefishes
Lophius americanus. Goosefish or Monkfish
Antennariidae. 42 species of frogfishes
Histrio histrio. Sargassumfish
Tetrabrachiidae. Monotypic Tetrabrachid Frogfish
Lophichthyidae. Monotypic, Lophichthyid Frogfish
Brachionichthyidae. 4 described species of handfishes
Chaunacidae. 14 species of coffinfishes or sea toads
Ogcocephalidae. 68 species of batfishes
Caulophrynidae. 5 species of fanfins
Neoceratiidae. Monotypic, Toothed Seadevil
Melanocetidae. 5 species of black seadevils
Himantolophidae. 18 species of footballfishes
Diceratiidae. 6 species of double anglers
Oneirodidae. About 62 species of dreamers
Thaumatichthyidae. 7 species of wolftrap anglers
Lasiognathus amphirhamphus. Wolftrap Angler
Centrophrynidae. Monotypic, Deepsea Anglerfish
Ceratiidae. 4 species of seadevils
Gigantactinidae. 22 species of whipnose anglers
Linophrynidae. 23 species of leftvents
Photocorynus spiniceps
Linophryne arborifera. Illuminated Netdevil

Order Pleuronectiformes. 14 families and 678
    species of flatfishes
    Psettodidae. 3 species of spiny turbots
    Citharidae. 6 species of largescale
      flounders
    Scophthalmidae. About 8 species of
      turbots
    Paralichthyidae. About 105 species of
      sand flounders
      Paralichthys dentatus. Summer
        Flounder
    Pleuronectidae. About 60 species of
      righteye flounders
      Hippoglossus. Genus of halibuts
      Pseudopleuronectes americanus. Winter
        Flounder
    Bothidae. About 140 species of lefteye
      flounders
    Paralichthodidae. Monotypic, Measles
      Flounder
    Poecilopsettidae. 20 species of bigeye
      flounders
    Rhombosoleidae. 19 species of
      rhombosoleids
    Achiropsettidae. 5 or 6 species of
      southern flounders
    Samaridae. About 20 species of crested
      flounders

    Achiridae. About 33 species of American
      soles
    Soleidae. About 130 species of soles
    Cynoglossidae. About 127 species of
      tonguefishes
Order Tetraodontiformes. 9 families with 357
    species
    Triacanthodidae. 21 species of spikefishes
    Triacanthidae. 7 species of triplespines
    Balistidae. About 40 species of
      triggerfishes
    Monacanthidae. About 102 species of
      filefishes
    Ostraciidae. 33 species of boxfishes and
      trunkfishes
      Triodontidae. Monotypic, Threetooth
        Puffer
    Tetraodontidae. About 130 species of
      puffers
      Canthigaster. 32 species of sharpnose
        puffers
      Tetraodon. Genus of freshwater puffers
    Diodontidae. 19 species of
      porcupinefishes and burrfishes
    Molidae. 4 species of ocean sunfishes
      Mola mola. Ocean Sunfish

Appendix B

Some Organizations That Promote Ichthyology and the Conservation of Fishes

Most of the organizations hold annual meetings and have websites that can be found by searching their name.

American Elasmobranch Society. Focuses on biology and conservation of sharks, skates, and rays.

American Fisheries Society. Publisher of *Transactions of the American Fisheries Society*, *North American Journal of Fisheries Management*, *North American Journal of Aquaculture*, *Journal of Aquatic Animal Health*, and the newsletter *Fisheries*.

American Institute of Fishery Research Biologists. Publishes a newsletter *Briefs*.

American Society of Ichthyologists and Herpetologists. Publisher of the journal *Copeia*.

Desert Fishes Council. Focuses on conservation of desert fishes of the U.S. and Mexico.

European Ichthyological Society. Sponsor of the European Congresses of Ichthyology.

Fishery Society of the British Isles. Publisher of the *Journal of Fish Biology*.

Ichthyological Society of Japan. Publisher of two journals, one in English *Ichthyological Research* and one in Japanese, *Japanese Journal of Ichthyology*.

North American Native Fishes Association. Focuses on capture, care, and conservation of native fishes. Publishes *American Currents*.

Southeastern Fishes Council. Publisher of *Southeastern Fishes Council Proceedings*.

Bibliography

Adams, S. M., ed. 1990. *Biological Indicators of Stress in Fish*. Bethesda, MD: American Fisheries Society Special Publication 8.

Allen, G. R. 1991. *Damselfishes of the World*. Melle, Germany: Mergus.

Allen, L. G., D. J. Pondella II, and M. H. Horn. 2006. *The Ecology of Marine Fishes. California and Adjacent Waters.* Berkeley, CA: University of California Press.

Arratia, G., B. G. Kapoor, M. Chardon, and R. Diogo, eds. 2003. *Catfishes.* vols. 1 and 2. Enfield, NH: Science Publishers.

Axelrod, H. R., C. W. Emmens, W. E. Burgess, N. Pronek, and G. S. Axelrod. 1996. *The Encyclopedia of Freshwater Tropical Fishes,* expanded ed. Neptune City, NJ: TFH Publications.

Barlow, G. W. 2000. *The Cichlid Fishes: Nature's Grand Experiment in Evolution*. Cambridge, MA: Perseus Publishing.

Barthem, R., and M. Goulding. 1997. *The Catfish Connection: Ecology, Migration, and Conservation of Amazon Predators*. New York: Columbia University Press.

Barton, M. 2006. *Bond's Biology of Fishes,* 3rd ed. Stamford, CT: Thomson Brooks/ Cole.

Becker, G. C. 1983. *Fishes of Wisconsin*. Madison, WI: University of Wisconsin Press.

Behnke, R. J. 2002. *Trout and Salmon of North America*. New York: The Free Press.

Bemis, W. E., W. W. Burggren, and N. E. Kemp, eds. 1987. *The Biology and Evolution of Lungfishes*. New York: Alan R. Liss.

Berra, T. M. 2007. *Freshwater Fish Distribution*. Chicago: University of Chicago Press.

Billard, R., and G. Lecointre. 2001. Biology and conservation of sturgeon and paddlefish. *Reviews in Fish Biology and Fisheries* 10:355–392.

Block, B., and E. Stevens, eds. 2001. *Tuna: Physiology, Ecology, and Evolution*. Fish Physiology, vol. 19. New York: Academic Press.

Böhlke, J. E., and C. C. G. Chaplin. 1993. *Fishes of the Bahamas and Adjacent Tropical Waters*. 2nd ed. Austin: University of Texas Press.

Bone, Q., N. B. Marshall, and J. H. S. Blaxter. 1995. *Biology of Fishes*. 2nd ed. London: Blackie Academic and Professional.

Boschung, H.T., Jr., and R. L. Mayden. 2004. *Fishes of Alabama*. Washington, DC: Smithsonian Institution Press.

Breder, C. M., Jr., and D. E. Rosen. 1966. *Modes of Reproduction in Fishes*. Neptune City, NJ: TFH Publications.

Burgess, W. E. 1978. *Butterflyfishes of the World. A Monograph of the Family Chaetodontidae*. Neptune City, NJ: TFH Publications.

Burgess, W. E. 1989. *An Atlas of Freshwater and Marine Catfishes: A Preliminary Survey of the Siluriformes*. Neptune City, NJ: TFH Publications.

Burgess, W. E., H. R. Axelrod, and R. E. Hunziker III. 1988. *Dr. Burgess's Atlas of Marine Aquarium Fishes*. Neptune City, NJ: TFH Publications.

Cailliet, G. M., M. S. Love, and A.W. Ebeling. 1986. *Fishes: A Field and Laboratory Manual on Their Structure, Identification, and Natural History*. Belmont, CA: Wadsworth Publishing.

Cohen, D. M. 1970. How many Recent fishes are there? *Proceedings of the California Academy of Sciences*, 4th series 38: 341–346.

Collette, B. B., and G. Klein-MacPhee, eds. 2002. *Bigelow and Schroeder's Fishes of the Gulf of Maine*. 3rd ed. Washington, DC: Smithsonian Institution Press.

Courtenay, W. R., Jr., and J. R. Stauffer, Jr., eds. 1984. *Distribution, Biology, and Management of Exotic Fishes*. Baltimore: Johns Hopkins University Press.

Courtenay, W.R., Jr., and J. D. Williams. 2004. Snakeheads (Pisces, Channidae): A biological synopsis and risk assessment. *U.S. Geological Survey Circular* 1251.

Diana, J. S. 2004. *Biology and Ecology of Fishes*. 2nd ed. Traverse City, MI: Biological Sciences Press.

Dolin, E. J. 2003. *Snakehead: A Fish out of Water*. Washington, DC: Smithsonian Books.

Douglas, N. H. 1974. *Freshwater Fishes of Louisiana*. Baton Rouge, LA: Claitor's Publication Division.

Eastman, J. T. 1993. *Antarctic Fish Biology: Evolution in a Unique Environment*. San Diego, CA: Academic Press.

Ellis, R. 2008. *Tuna: A Love Story*. New York: Alfred A. Knopf.

Eschmeyer W. N. 1990. *Catalog of the Genera of Recent Fishes*. San Francisco: California Academy of Sciences.

Eschmeyer, W. N., E. S. Herald, and H. Hammann. 1983. *A Field Guide to Pacific Coast Fishes of North America*. Peterson Field Guide Series. Boston: Houghton Mifflin.

Etnier, D. A., and W. C. Starnes. 1993. *The Fishes of Tennessee*. Knoxville: University of Tennessee Press.

Evans, D. H., and J. B. Claiborne. 2006. *The Physiology of Fishes*. 3rd ed. Boca Raton, FL: CRC, Taylor & Francis.

FAO, 2007. *The State of World Fisheries and Aquaculture (SOFIA) 2006*. Rome: FAO.

Franklin, H. B. 2008. *The Most Important Fish in the Sea: Menhaden and America*. South Beach, OR: Shearwater Press.

Fryer, G., and T. D. Iles. 1972. *The Cichlid Fishes of the Great Lakes of Africa*. Edinburgh: Oliver and Boyd.

Fukuchi, M., H. J. Marchant, and B. Nagase. 2006. *Antarctic Fishes*. Baltimore: Johns Hopkins University Press.

Fuller, P. L., L. G. Nico, and J. D. Williams. 1999. *Nonindigenous Fishes Introduced into Inland Waters of the United States*. Bethesda, MD: American Fisheries Society Special Publication 27.

Gage, J. D. and P. A. Tyler. 1991. *Deep-sea Biology: A Natural History of Organisms at the Deep-sea Floor*. Cambridge: Cambridge University Press.

Godin, J. J., ed. 1997. *Behavioural Ecology of Teleost Fishes*. Oxford: Oxford University Press.

Goldschmidt, T. 1996. *Darwin's Dreampond: Drama in Lake Victoria*, translated by S. Marx-Macdonald. Cambridge, MA: The MIT Press.

Goulding, M. 1980. *The Fishes and the Forest: Explorations in Amazonian Natural History*. Berkeley: University of California Press.

Graham, J. B. 1997. *Air-breathing Fishes: Evolution, Diversity, and Adaptation*. San Diego, CA: Academic Press.

Greenwood, P. H. 1981. *The Haplochromine Fishes of the East African Lakes*. Ithaca, NY: Cornell University Press.

Groot, C., and L. Margolis, eds. 1991. *Pacific Salmon Life Histories*. Vancouver: University of British Columbia Press.

Halverson, A. 2010. *An Entirely Synthetic Fish: How Rainbow Trout Beguiled America and Overran the World*. New Haven, CT: Yale University Press.

Hart, J. L. 1980. *Pacific Fishes of Canada*. Ottawa: Fisheries Research Board of Canada Bulletin 180.

Hart, P. J. B., and J. D. Reynolds, eds. 2002. *Handbook of Fish Biology and Fisheries*. vol. 1. *Fish Biology*. vol. 2. *Fisheries*. Oxford: Blackwell Science.

Hartel, K. E., D. B. Halliwell, and A. E. Launer. 2002. *Inland Fishes of Massachusetts*. Lincoln: Massachusetts Audubon Society.

Helfman, G. S. 2007. *Fish Conservation: A Guide to Understanding and Restoring Global Aquatic Biodiversity and Fishery Resources*. Washington, DC: Island Press.

Helfman, G. S., B. B. Collette, D. E. Facey, and B. Bowen. 2009. *The Diversity of Fishes: Biology, Evolution, and Ecology*. 2nd ed. Oxford: Wiley-Blackwell.

Hersey, J. 1988. *Blues*. New York: Random House.

Hocutt, C. H., and E. O. Wiley, eds. 1986. *The Zoogeography of North American Freshwater Fishes*. New York: Wiley.

Horn, M. H. 1989. Biology of marine herbivorous fishes. *Oceanography Marine Biology, Annual Review* 27:167–272.

Horn, M. H., K. L. M. Martin, and M. A. Chotkoski, eds. 1999. *Intertidal Fishes: Life in Two Worlds*. San Diego, CA: Academic Press.

IUCN 2010. The IUCN Red List of Threatened Species, 2010. www.iucnredlist.org.

Jelks, H. L., S. J. Walsh, N. M. Burkhead, et al. 2008. Conservation status of imperiled North American freshwater and diadromous fishes. *Fisheries* 33(8): 372–407.

Jenkins, R. E., and N. M. Burkhead. 1993. *Freshwater Fishes of Virginia*. Bethesda, MD: American Fisheries Society.

Johannes, R. E. 1981. *Words of the Lagoon: Fishing and Marine Lore in the Palau District of Micronesia*. Berkeley, CA: University of California Press.

Johnson, D. G. D., J. R. Paxton, et al. 2009. Deep-sea mystery solved: Astonishing larval transformations and extreme sexual dimorphism unite three fish families. *Biology Letters* 5:235–239.

Keenleyside, M. H. A., ed. 1991. *Cichlid Fishes: Behaviour, Ecology and Evolution*. London: Chapman & Hall.

Knecht, G. B. 2007. *Hooked: Pirates, Poaching, and the Perfect Fish*. Emmaus, PA: Rodale Books.

Kottelat, M., R. Britz, T. H. Hui, and K.-E. Witte. 2006. *Paedocypris*, a new genus of Southeast Asian cyprinid fish with a remarkable sexual dimorphism, comprises the world's smallest vertebrate. Proceedings of the Royal Society (London) B 273: 895–899.

Kottelat, M., and J. Freyhof. 2007. *Handbook of European Freshwater Fishes*. Cornol, Switzerland: Publications Kottelat.

Kurlansky, M. 1997. *Cod: A Biography of the Fish That Changed the World*. New York: Walker and Co.

Ladich, F., P. C. Shaun, P. Moller, and B. G. Kapoor, eds. 2006. *Communication in Fishes*. Enfield, NH: Science Publishers.

Lamboj, A. 2004. *The Cichlid Fishes of Western Africa*. Bornheim, Germany: Birgit Schmettkamp Verlag.

Lasker, R., ed. 1981. *Marine Fish Larvae*. Seattle: Washington Sea Grant Publications.

Lee, D. S., C. R. Gilbert, C. H. Hocutt, R. E. Jenkins, D. E. McAllister, and J. R. Stauffer, Jr. 1980. *Atlas of North American Fresh Water Fishes*. Raleigh: North Carolina State Museum of Natural History.

Lévêque, C. 1997. *Biodiversity Dynamics and Conservation: The Freshwater Fish of Tropical Africa*. Cambridge: Cambridge University Press.

Lever, C. 1996. *Naturalized Fishes of the World*. San Diego, CA: Academic Press.

Lichatowich, J. 1999. *Salmon without Rivers: A History of the Pacific Salmon Crisis*. Covelo, CA: Island Press.

Long, J. A. 2010. *The Rise of Fishes*, 2nd ed. Baltimore, MD: Johns Hopkins University Press.

Love, M. S., M. Yoklavich, and L. K. Thorsteinson. 2002. *The Rockfishes of the Northeast Pacific*. Berkeley, CA: University of California Press.

Lowe-McConnell, R. H. 1987. *Ecological Studies in Tropical Fish Communities*. Cambridge: Cambridge University Press.

Lucas, M. C., and E. Baras. 2001. *Migration of Freshwater Fishes*. Oxford: Blackwell.

Maisey, J. G. 1996. *Discovering Fossil Fishes*. New York: Henry Holt and Co.

Marshall, N. B. 1971. *Explorations in the Life of Fishes*. Cambridge, MA: Harvard University Press.

Matthews, W. J. 1998. *Patterns in Freshwater Fish Ecology*. New York: Chapman & Hall.

McDowall, R. M. 1988. *Diadromy in Fishes*. London: Croom Helm.

McEachran, J. D., and J. D. Fechhelm. 1998–2005. *Fishes of the Gulf of Mexico*. vol. 1, *Myxiniformes to Gasterosteiformes*. vol. 2, *Scorpaeniformes to Tetraodontiformes*. Austin: University of Texas Press.

McPhee, J. 2002. *The Founding Fish*. New York: Farrar, Straus and Giroux.

Mecklenburg, C. W., T. A. Mecklenburg, and L. K. Thorsteinson. 2002. *Fishes of Alaska*. Bethesda, MD: American Fisheries Society.

Meffe, G. K., and F. F. Snelson, Jr., eds. 1989. *Ecology and Evolution of Livebearing Fishes* (Poeciliidae). Englewood Cliffs, NJ: Prentice-Hall.

Miller, R. J., and H .W. Robison. 2004. *Fishes of Oklahoma*. Norman: University of Oklahoma Press.

Miller, R. R. 2005. *Freshwater Fishes of México*. Chicago: University of Chicago Press.

Montgomery, D. R. 2003. *King of Fish: The Thousand Year Run of Salmon*. Boulder, CO: Westview Press.

Moyle, P. B. 2002. *Inland Fishes of California*. revised ed. Berkeley, CA: University of California Press.

Moyle, P. B., and J. J. Cech. 2004. *Fishes: An Introduction to Ichthyology.* 5th ed. Upper Saddle River, NJ: Prentice-Hall.

Murdy, E. O., R. S. Birdsong, and J. A. Musick. 1997. *Fishes of Chesapeake Bay.* Washington, DC: Smithsonian Institution Press.

Nelson, J. S. 2006. *Fishes of the World.* 4th ed. Hoboken, NJ: John Wiley & Sons.

Nelson, J. S., chair. 2004. *Common and Scientific Names of Fishes from the United States, Canada, and Mexico,* 6th ed. Bethesda, MD: American Fisheries Society Special Publication 29.

Nelson, J. S., H.-P. Schultze, and M. V. H. Wilson, eds. 2010. *Origin and Phylogenetic Interrelationships of Teleosts.* Munich, Germany: Friedrich Pfeil.

Northcutt, R.G., and R. E. Davis, eds. 1983. *Fish Neurobiology.* Ann Arbor: University of Michigan Press, 2 vols.

Ostrander, G. K., ed. 2000. *The Laboratory Fish.* San Diego, CA: Academic Press.

Page, L. M. 1983. *Handbook of Darters.* Neptune City, NJ: TFH Publications.

Page, L. M., and B. M. Burr. 1991. *A Field Guide to Freshwater Fishes: North America North of Mexico.* Peterson Field Guide Series. Boston: Houghton Mifflin.

Paxton, J. R., and W. N. Eschmeyer, eds. 1998. *Encyclopedia of Fishes.* 2nd ed. San Diego, CA: Academic Press.

Perry, S. F., and B. L. Tufts. 1998. *Fish Respiration.* vol. 17, *Fish Physiology.* New York: Academic Press.

Pflieger, W. L. 1997. *The Fishes of Missouri.* Jefferson City: Missouri Department of Conservation.

Pietsch, T. W. 2009. *Oceanic anglerfishes: Extraordinary diversity in the deep sea.* Berkeley, CA: University of California Press.

Pietsch, T. W. 2010. *Samuel Fallours: Tropical Fishes of the East Indies,* ed. by Petra Lamers. Cologne, Germany: Schutze Taschen.

Pietsch, T. W., and D. B. Grobecker. 1987. *Frogfishes of the World: Systematics, Zoogeography, and Behavioral Ecology.* Stanford, CA: Stanford University Press.

Pitcher, T. J., ed. 1993. *The Behaviour of Teleost Fishes.* 2nd ed. London: Chapman & Hall.

Potts, G. W., and R .J. Wootton, eds. 1984. *Fish Reproduction: Strategies and Tactics.* London: Academic Press.

Proudlove, G. S. 2006. *Subterranean Fishes of the World: An Account of the Subterranean (Hypogean) Fishes Described up to 2003 with a Bibliography 1541–2004.* Moulis, France: International Society for Subterranean Biology.

Quinn, T. P. 2005. *The Behavior and Ecology of Pacific Salmon and Trout.* Seattle: University of Washington Press.

Randall, D. J., and A. P. Farrell, eds. 1997. *Deep-sea Fishes.* San Diego, CA: Academic Press.

Randall, J. E. 2007. *Reef and Shore Fishes of the Hawaiian Islands.* Honolulu: Sea Grant College Program, University of Hawai'i.

Reebs, S. 2001. *Fish Behavior in the Aquarium and in the Wild.* Ithaca, NY: Cornell University Press.

Robins, C. R., and G. C. Ray. 1986. *A Field Guide to Atlantic Coast Fishes of North America.* Peterson Field Guide Series. Boston: Houghton Mifflin.

Robison, H. W., and T. M. Buchanan. 1988. *Fishes of Arkansas*. Fayetteville: University of Arkansas Press.

Roberts, C. 2008. *The Unnatural History of the Sea*. South Beach, OR: Shearwater Press.

Sale, P. F., ed. 2002. *Coral Reef Fishes: Dynamics and Diversity in a Complex Ecosystem*. San Diego, CA: Academic Press.

Schreck, C. B., and P. B. Moyle, eds. 1990. *Methods for Fish Biology*. Bethesda, MD: American Fisheries Society.

Scott, W. B., and E. J. Crossman. 1973. *Freshwater Fishes of Canada*. Fisheries Research Board of Canada Bulletin 184.

Scott, W. B., and M. G. Scott. 1988. *Atlantic Fishes of Canada*. Canadian Bulletin of Fisheries and Aquatic Science, no. 219. Toronto: University of Toronto Press.

Sinderman, C. J. 1990. *Principal Diseases of Marine Fish and Shellfish*. 2nd ed. Vol. 1. *Diseases of Marine Fish*. San Diego, CA: Academic Press.

Sloman, K. A., R. W. Wilson, and S. Balshine. 2005. *Behaviour and Physiology of Fish*. Fish Physiology, vol. 24. New York: Academic Press.

Smith, C. L. 1985. *The Inland Fishes of New York State*. Albany: New York State Department of Environmental Conservation.

Smith, C. L. 1997. *National Audubon Society Field Guide to Tropical Marine fishes of the Caribbean, the Gulf of Mexico, Florida, the Bahamas, and Bermuda*. New York: Alfred A. Knopf.

Smith-Vaniz, W. F., B. B. Collette, and B. E. Luckhurst. 1999. *Fishes of Bermuda: History, Zoogeography, Annotated Checklist, and Identification Keys*. American Society of Ichthyologists and Herpetologists Special Publication 4.

Sneddon, L. U., V. A. Braithwaite, and M. J. Gentle. 2003. Do fishes have nociceptors? Evidence for the evolution of a vertebrate sensory system. *Proceedings of the Royal Society London B* 270:1115–1121.

Spotte, S. 2002. *Candiru: Life and Legend of the Bloodsucking Catfishes*. Berkeley, CA: Creative Arts Book Co.

Stauffer, J. R., Jr., J. M. Boltz, and L. R. White. 1995. *The Fishes of West Virginia*. Proceedings of the Academy of Natural Sciences of Philadelphia 146: 1–389.

Tesch, F.-W., R. J. White, and J. Thorpe. 2003. *The Eel*, translated by J. Greenwood, 5th ed. Hoboken, NJ: Wiley-Blackwell.

Thresher, R. E. 1984. *Reproduction in Reef Fishes*. Neptune City, NJ: TFH Publications.

Trautman, M. N. B. 1981. *The Fishes of Ohio*, rev. ed. Columbus: Ohio State University Press.

Uribe, M. C. A., and H. J. Grier, eds. 2005. *Viviparous Fishes*. Homestead, FL: New Life Publications.

Val, A. L., V. M. F. De Almeida-Val, and D. J. Randall. 2005. *The Physiology of Tropical Fishes*, vol. 21, *Fish Physiology*. New York: Academic Press.

Webb, P. W., and D. Weihs. 1983. *Fish Biomechanics*. New York: Praeger.

Weber, M. 1913. Susswasserfische aus Niederlandisch sud-und nord-Neu-Guinea. In: Nova Guinea. Resultats de L'Expedition Scientifique Neerlandaise a la Nouvelle-Guinee en 1907 et 1909. Zoologie. 9:513–613, Plates 12–14., Leiden, The Netherlands: Brill.

Weinberg, S. 2000. *A Fish Caught in Time. The Search for the Coelacanth*. New York: HarperCollins.

Westerfield, M. 2000. *The Zebrafish Book. A Guide for the Laboratory Use of Zebrafish (Danio rerio)*. 4th ed. Eugene: University of Oregon Press.

Wiley, E. O., and G. D. Johnson. 2010. A teleost classification based on monophyletic groups. Pp 123–182 in J. S. Nelson, H.-P. Schultze, and M. V. H. Wilson, eds. 2010. *Origin and Phylogenetic Interrelationships of Teleosts*. Munich, Germany: Friedrich Pfeil.

Wootton, R. J. 1984. *A Functional Biology of Sticklebacks*. London: Croom Helm.

Wootton, R. J. 1998. *Ecology of Teleost Fishes*. 2nd ed. Boston, MA: Kluwer Academic Publishers.

Wourms, J. P., B. D. Grove, and J. Lombardi. 1988. The maternal embryonic relationship in viviparous fishes. In: Hoar and Randall, eds. *Fish Physiology*, vol. 11b, San Diego: Academic Press.

Index

Page numbers in **bold** refer to illustrations.

Actinopterygii (ray-finned fishes), 7, 8

adult fishes, 28, 33, 34, 80, **81**, 148

age, 28, 33–34, 82, 85–86

Agnatha, 2

Alewife (*Alosa pseudoharengus*, Clupeidae), 62, 111–12

algae, 38, 39, 41, 44, 72, 87, 116; blooms of, 4, 70, 123; and camouflage, 29, 55; as food, 14, 71, 93–94

American Marinelife Dealers Association's Ecolist, 104, 126

anadromous fishes, 20

anatomy, 12–17, 18, 146; anus, 13; blood, 17, 51, 66, 69, 75; bones, 8, 12, **13**, 16; brain, 24, 51, 52, 63–64, 70, 106; breathing, 17–18, 69; and buoyancy, 18, **19**; caudal fin, 14; caudal muscles, 23; digestive tract, 66; ears, 85, 97; and ectotherms and endotherms, 17; and energy and nutrients, 71–72; esophagus, 15; gas bladders, 3, 18, **19**, 52, 67, 97; intestinal tract, 93, 114; jaws, 15, 87, 89, 91, 93–94, 95, 102; kidneys, 20; mouths, 66, 83, 84, 87, 89, 90, 91, 94, 95; and pain, 106; and poisonous spines, 55, 57; and reversal of sex, 74, 75, 82–83, 120; and sense of smell, 63, 67, 98; and senses, 96–100; and sexual dimorphism, 25; stomach, 91, 93, 94; throat, 14, 66. *See also* eyes; gills; reproduction/breeding; scales; teeth

anchovies (Engraulidae), 2, 8, 26, 43, 71, 121, 144

Anemonefishes (*Amphiprion, Premnas*, Pomacentridae), 38, 40–41, 72, 73, 75, 82–83, 146

angelfishes: Pomacanthidae, 34, 41, 72, 88; *Pterophyllum*, Cichlidae, **3**, 146

anglerfishes: Ceratioidei, 8, 14, 25, 65, 75, 89, 99–100, 133; *Photocorynus spiniceps*, Linophrynidae, **76**

aquariums, **3**, 28, 35, 70, 88, 103–5, 121, 122, 126

Arapaima/Pirarucu (*Arapaima gigas*, Osteoglossidae), 10, **12**, 59, 62

archerfish (*Toxotes*, Toxotidae), 102, **102**

Atlantic Cod (*Gadus morhua*, Gadidae), 4, 78, 142

Atlantic Salmon (*Salmo salar*, Salmonidae), 35, 136, 145

bacteria, 24, 69, 93, 114

baitfishes, 26, 43, 47, 71

barbs (Cyprinidae), 2, 103, 146; *Puntius*, 88; African cyprinids (*Barbus*), 35

barracudas (Sphyraenidae), 14, 36, 42, 58, 89, **90**, 91, 96; Great (*Sphyraena barracuda*, Sphyraenidae), **35**, 36, 39, **45**, 46, 115; and humans, 108, 112; and territoriality, 37

basses, 13, 110. *See also* basses, types of; black basses and sunfishes

basses, types of: Kelp (*Paralabrax clathratus*, Serranidae), 82; Peacock (*Cichla ocellaris*), 90, 111; Rock, (*Ambloplites*, Centrarchidae), 32. *See also* basses; Largemouth Bass; Smallmouth Bass; Striped Bass

batfishes (Lophiiformes), 99; Platacidae, 34

behavior, 37–60; and avoidance of predators, 53–60; and biting, 38, 47–48; and cannibalism, 75; cleaning, 41–42, 48, 72, 73, 88, 110; and coloration, 28, 30; commensalism, 40; and communication, 23, 52–53; coprophagy, 88; and culture, 135–39, 143; and fighting, 46–47; and interspecific interactions, 39–41; and learning, 48–51; and migration, 49, 61–63, 69, 72, 119; and navigation, 23, 24, 63; and play, 51–52; and resting, 20, 28, 29, 38, 43, 49, 61, 67; and salt water–fresh water movement, 5, 20; of scavengers, 48, 100–101; schooling, 42–46, 52, 53, 54, **59**; social, 37–46, 82; and sounds, 18, 60, 97–98, 124; symbiosis, 39–41, 72–73; and talking, 52–53; territoriality, 37–39, 46, 48, 49, 52

bichirs, 8, 59, 66, 94

bigeyes (Priacanthidae), 20, 30, 43, 96

bignose fishes (Megalomycteridae), 147–48

billfishes (Istiophoridae), 10, 62, 71, 90–91, 95, 146

bioluminescence, 24–25

Bitterlings (*Rhodeus*, Cyprinidae), 78–79

black basses and sunfishes (*Micropterus*, Centrarchidae), 2, 37, 74, 91, 96, 145. *See* ˌ basses; basses, types of; *individual spec*

Black Crappie (*Pomoxis nigromaculatus*, C ˌ trarchidae), 43, 96

blacklip pearl oysters (*Pinctada*), 40, **4]**

Blacksmith (*Chromis punctipinnis*, Pon ˌ dae), 71–72

blennies (Blenniidae), 32, 37, 53, 5ᴶ saber-tooth (*Aspidontus, Plagio* 47–48, 88

blowfishes (Tetraodontidae), 14

171

Bluefish (*Pomatomus saltatrix*, Pomatomidae), 4, 47, **48**, 56, 58, 91, 112, 142

Bluegill Sunfish (*Lepomis macrochirus*, Centrarchidae), **38**, 42, 57, 87, 89, 91, 110–11

bodies: shape of, 89-90, 92, 146; size of, 10-11, 38, 76, 82; temperature of, 17

Bonefish (*Albula vulpes*), 8, 145

bony fishes (Osteichthyes/Teleostomi), 2, 6–7, 12, 13, **13**, 22, 67, 75, 94; feeding by, 93; longest, 10; major groups of, 7–8; scales of, 15; teeth of, 14

Bowfin (*Amia calva*, Amiidae), 8, **9**, 14, 77, 81, 94, 95, 111; and dermal armor, 59; injury from, 112

Burbot *(Lota lota)*, 68, 78

butterflyfishes (*Chaetodon*, Chaetodontidae), 20, 37, 41, 50, 62, 72, 88, 136; and coloration, 30, 32, 34; colored, 28

camouflage, 29, 54, 100. *See also* coloration

Candiru (*Vandellia cirrhosa*, Trichomycteridae), 113, **114**

cardinalfishes (Apogonidae), 20, 40, 43, 54, 62, 81, 96

carps (Cyprinidae), 15, 103, 121, 145, 146; Asian, 56, 78; Chinese, 111; Common (*Cyprinus carpio*, Cyprinidae), 32, 68, 87, 111; Julien's Golden (*Probarbus jullieni*, Cyprinidae), 118; Koi, 109

cartilaginous fishes (Chondrichthyes), 2, 12, 15, 22

catadromous fishes, 20

catfishes (Siluriformes), 3, 8, 13, 57, 61, 103, 112, 133, 146; and alarm substances, 59; and caves, 67; and dermal armor, 59; effect on environment, 71; and electrosensory organs, 98; feeding by, 93; and fish farms, 121; and hearing, 97; and parenting, 84; and relatedness of fish in nest, 80; and reproduction, 74; scavenging by, 101; and sex determination, 82; and talking, 53; and taste buds, 98; and territoriality, 37. *See also* catfishes, types of

catfishes, types of: Ariidae, 58; armored (*Corydoras*, Callichthyidae), 79, 146; bagrid, 81; bullhead (*Ameiurus*, Ictaluridae), 87, 111; Callichthyid (*Corydoras semiaquilus*), **79**; Chacidae, 100; Channel (*Ictalurus punctatus*, Ictaluridae), 145; Clariidae, 67; Cuckoo (*Synodontis multipunctata*, Mochokidae), 84; eel-tailed (Australian tandan; *Plotosus*, Plotosidae), 57, **58**; electric (*Malapterurus*, Malapteruridae), 22, 23, 112; Ictaluridae, 16; Loricariidae, 16, **105**; madtom (*Noturus*, Ictaluridae), 77; Mekong Giant (*Pangasianodon gigas*, Pangasiidae), **64**, 118; Mochokidae, 31; pleco, 104; suckermouth (Loricariidae), 146; upside-down (*Synodontis nigriventris*, Mochokidae), 22, 31,

99; Walking (*Clarias batrachus*, Clariidae), 17–18, 67, 104, 111. *See also* Candiru; sea catfishes

cavefishes (Amblyopsidae), 67–68, 97, 118

cephalochordate, 8

Cero Mackerel (*Scomberomorus regalis*), **28**

Chain Pickerel, 68

Characiformes, 2–3

characins (Characiformes), 8, 13, 14, 19, 67, 71, 93, 97; and migration, 62; replacement dentition of, 95; and talking, 53

chimaeras, 2, 12, 96

Chinook Salmon (*Oncorhynchus tshawytscha*, Salmonidae), 2, 63, **85**, 138

Chondrichthyes, 2

Chordata, 6

chubs *(Nocomis)*, 72, 81

cichlids (Cichlidae), 3, **3**, 14, 15, 31, 64, 82, **88**, 90, 146; and alarm substances, 59; and aquarium keeping, 103; and coloration, 34, 35; *Docimodus johnstoni*, 48; effect on environment, 71; and egg laying, 79; feeding by, 87, 88, 89, 93, 101; *Lamprologus, Neolamprologus, Julidochromis*, 84; and luring, 100; and parenting, 84; pipette mouths of, 89; and predation, 90; and relatedness of fish in nest, 80, 81; *Rhamphochromis*, 90; and talking, 53; and territoriality, 37; *Tyrannochronis nigriventer*, 31. *See also* individual species

ciguatera, 115–16

climate change, 118–20

clinids, 71–72

Clupeomorpha, 8

cods (Gadidae), 2, 8, 13, 65, 76, 78, 80, 121; *Gadus morhua*, 77; Haddock and Pollock, 144; and hearing, 97; and talking, 53. *See also* Haddock; Pollock; *individual species*

coelacanths (Latimeriidae), 7, 12, 62, 75, 77, 78, 94, 96; African (*Latimeria chalumnae*, Latimeriidae), **8**, 118; and electrosensory organs, 98; fossil, 15; Indonesian (*L. manadoensis*), 118. *See also individual species*

coloration: and camouflage, 29; and cavefishes, 67, 68; changes in, 28, 33–34; and cleaning, 42; control over, 30; and countershading, 29–31, 34, 56; differences among, 27–29, 146; of eyes, 32–33; and fighting, 46–47; geographic variation in, 35–36; parr marks, **35**; and predator avoidance, 54–56; reasons for, 29–32; seasonal changes of, 34; silvery, 26–27, 29, 34, **35**, 54, 56, 96; split head, 32; and ultraviolet (UV) light, 28–29; warning, 57. *See also* camouflage

Convention on International Trade in Endangered Species (CITES), 118

coral reefs, 20, 43, 65, 74, 78, 88, 101; and climate change, 119; and coloration, 28, 34, **35**; and corals, 5, 40, 72, 93–94, 119;

dynamiting in, 122; ecology of, 71, 72; and feeding, 88, 96; and migration, 61, 62; and territoriality, 37–38. *See also* reef fishes; reefs

croakers (*Bairdiella*, Sciaenidae), 52, 61, 65, 124, 144

crocodilefishes, 29, 55

Cyprinodontiformes, 66; *Nothobranchius, Aphyosemion*, 78

damselfishes (Pomacentridae), 3, 20, 28, 37–38, 41, 88, 146; Beau Gregory (*Stegastes leucostictus*, Pomacentridae), 29, 39, 49; biting by, 48; blue-colored, 36; and coloration, 34; and coprophagy, 88; and ecosystem, 72; and environment, 71; feeding by, 93; and intelligence, 50; and migration, 61, 62; mouths of, 89; and predators, 54; and schooling, 44; and sex reversal, 74; and spawning season, 78; and talking, 53; and territoriality, 39

darters (Percidae), 21, 29, 53, 55, 59, 81, 118; and coloration, 30, 34; *Percina, Etheostoma*, 30

desert pupfishes (Cyprinodontidae), 66

detritivores, 89

Devils Hole Pupfish (*Cyprinodon diabolis*), 66, **67**

diadromous fishes, 20, **21**

die-offs, 123

dogfish (Squalidae), 24, 65

dolphinfishes. *See* Mahi-mahi/Dorado/Dolphinfish

dories (*Zeus*), 91

dragonfishes: Stomiiformes, 24–25, 65; suborder Notothenioidei, 69

drums (Sciaenidae), 3, 34, 52, 53, 61, 72, 101, 144

durophages, 101

ears. *See* anatomy: ears; hearing

eels (*Anguilla*, Anguillidae), 8, 14, 15, 20, 58, 61, 65, 80, 98, 101, 122, 133, 135; and coloration, 33, 34; freshwater, 37, 80, 82, 91

eels, types of: American (*Anguilla rostrata*), 34, **63, 92;** Asian Swamp, 111; conger (Congridae), **55**, 82, 91; cusk (Ophidiidae), 5, 65; European (*Anguilla anguilla*), 82, 133; Giant Cusk (*Spectrunculus grandis*), **66;** gulper (Eurypharyngidae), 100; snake (Ophichthidae), 91, 100, 104; swamp (Synbranchidae), 66, 91. *See also* Electric Eel; moray eels

eggs, 11, 38, 39, 66, 74, 81, 83; laying of, 75–76, 78–79. *See also* reproduction/breeding

Electric Eel (*Electrophorus electricus*, Gymnotidae), 23, 24, **24,** 66, 112

electric fishes, 22–24, 47, 51, 98–99. *See also* catfishes, types of: electric; knifefishes: electric; *individual species*

electric torpedo rays (Torpedinidae), 22, 23, 133, 134

electrosensory organs, 98–99

Elephantfishes (Mormyridae), 62, 99; African (Osteoglossiformes), 22, 24; *Gnathonemus petersii*, 51–52, **52**

Elopomorpha, 8

endangered species, 117–18

environment, 71–73, 82, 118–20

epipelagic fishes, **21**

Esociformes, 8, 14, 68

estivation, 19

Euteleostei, 8

evolution, 1, 6, 8–9, 71, 77

eyes, 20, 23, 24, 32–33, 67, 68; anatomy of, 21, 96; and feeding, 96–97; as food, 89; sight, 21, 29, 96-97. *See also* anatomy

farming (aquaculture), 121, 122, 144, 145

feces, 41, 71–72, 88, 94

feeding, 21, 40, 43, 46, 49, 61–62, 69; and luring, 25, 89, 99–100; and spin-feeders, 91–93; and territoriality, 37–38, 39. *See also* food; predators

females, 34, 39, 53, 74–75, 80, 81; and parenting, 84; and sex determination, 82–83

fighting, 28, 46–47, 74. *See also* predators

fin biters, 87, 88

fin rays (lepidotrichia), 13

fins, 1, 7, 13–14, 22, 42, 57, 75, 85; and coloration, 27; differences among, 146; and flying, 56; and rotational feeders, 92

fish: classification of, 5–8; importance of, 2–5; jawless, 2, 12; life cycles of, 5; life span of, 11–12; metabolism of, 17; speed and extent of growth of, 84–85; terminology for, 1–2

fishing, 45, 110, 111, 121–22, 125; lures for, 96, 98, 135; for sport, 2, 110, 121, 144, 145

fish kills, 107, 123, 124

fish leather, 122, **124,** 135

flashlight fishes (Anomalopidae), 24, 25

flatfishes and flounders (order Pleuronectiformes), 2, 29, 54, 55, 65, 69, 76, 89, 121; knowledge of, 144; Pleuronectidae, Paralichthyidae, 144; and sex determination, 82

flounder. *See* flatfishes and flounders

flyingfishes (Exocoetidae), 21–22, 56–57, 78; *Exocoetus*, **23**

food, 14, 21, 71, 89, 93, 101–2; chewing of, 94–95; and feeding by humans, 108–9: finding of, 96–100; for humans, 2, 4; of, 87–94; and luring, 99–100; and feeding fishes, 96; scavenging for, and spin-feeders, 91–93. *See also* fe predators

Four-eyed Fish (*Anableps*, Anablepid 96–97, **98**

French Grunt (*Haemulon flavolinea* mulidae), 46, 49

freshwater fishes, 5, 19, **21**, 34, 115, 118
frogfishes: Antennariidae, 37, 89, 91;
    Lophiiformes, 99

galaxiids, 59, 80
gars (Lepisosteidae), 8, 59, 80, 81, 89, **90**, 94,
    112; *Lepisosteus*, Lepisosteidae, 111
gas bladders. *See* anatomy: gas bladders
genetics, 82, 120, 146
geologic time periods, 8-9
gibberfishes (*Gibberichthys pumilus*, Gibberich-
    thyidae), 80, **81**
gills, 18, 19, 20, 23, 113, 125; cleaning of, 40, 42,
    88; and gill rakers, 87, 94; and respiration, 1,
    17, 18, 66, 67; and sickness, 70, 71, 120, **121**
global warming. *See* climate change
Gnathostomata (jawed vertebrates), 6
goatfishes (Mullidae), 98, **99**
gobies (Gobiidae), 3, 10, 29, 37, 40, 53, 59, 71;
    bumblebee (*Brachygobius*, Gobiidae), 88;
    dwarf (*Trimmaton nanus*), 10–11; *Gobiodon*,
    Gobiidae, 57; neon (*Gobiosoma*), 42, 88;
    pygmy coral reef (*Eviota sigillata*), 11; replace-
    ment dentition of, 95; Round, 111; and sex
    determination, 82; and sex reversal, 74; and
    shrimps, 72, 73; and spawning season, 78
Golden Dragonfish (*Scleropages formosus*, Osteo-
    glossidae), 118
Goldfish (*Carassius auratus*, Cyprinidae), 44,
    106, 111, 134
Goosefish (*Lophius americanus*, Lophiidae), 14,
    99, 112
gouramis (Anabantoidei), 3, 18, 67, 103, 146
groupers, 42, 44, 45, 53, 115, 126, 145; Giant
    (*Epinephelus lanceolatus*), **92**; Nassau,
    (*E. striatus*, Serranidae), 62, 82
grunts (Haemulidae), 20, 43, **50**, 52, 53, 61, 72,
    144; *Haemulon*, 143. *See also* French Grunt
gunnels (Pholidae), 55, 91
guppies (*Poecilia reticulata*, Poeciliidae), 35, 111
gyotaku, 135

habitats, 5, 19–20, **21**, 29, 64, 65–66, 67–68, 71;
    and coloration, 30, 33–34; desert, 66–67;
    disturbance to, 118; health of, 5; loss of,
    119; and migration, 61, 62; and mirror-
    sidedness, 26
Haddock, 4, 76, 144
hagfishes (Myxinidae), 2, 12, 37, 48, 58, 93, 94,
    101, 122
halfbeaks (*Dermogenys, Nomorhamphus, Hemir-
    hamphodon*; Zenarchopteridae), 56, 76, 93
halibuts (*Hippoglossus*), 47, 112, 144
hamlets (*Hypoplectrus*, Serranidae), 53, 75
hatchetfishes; freshwater (Gasteropelecidae),
    22, 57; marine, (Sternoptychidae), 65, 100
Hawkfishes (Cirrhitidae), 32
headlight fishes (*Diaphus*, Myctophidae), 25
hearing, 46, 97–98

herbivores, 71, 88, 89, 93
herrings (Clupeidae), 2, 8, 13; Atlantic (*Clupea
    harengus*), 7, 53, 62; and coloration, 33; egg
    laying by, 78; feeding by, 93; and fisher-
    ies, 121; and food chains, 71; hearing of,
    97; knowledge of, 144; and migration, 61;
    mirror-sidedness of, 26; Pacific (*Clupea,
    Clupeidae*), 53; and pipette mouths, 89;
    river (*Alosa*), 20; and schooling, 43; and
    talking, 53
humans, 2, 3, 4, 7, 112–13, 114–16, 120, 121;
    boats of, 123–24

icefishes, 69
ichthyology, 144, 148
intelligence, 48–51
International Union for the Conservation of
    Nature (IUCN), 117
introduced species, 67, 68, 111, 118

jacks (Carangidae), 42, **58**
jellyfishes, 40
John Dory / St. Pierre Fish (*Zeus faber*, Zeidae),
    132
juveniles, 28, 33, 34, 63, 72, 77; names of, 80.
    See also reproduction/breeding

Kasidoridae, 80, **81**
kelpbeds, 42, 61, 71, 72, 96
killifishes (Cyprinodontiformes), 11, 34, 56,
    59, 66, 78, 82, 103, 146; Pike (Poeciliidae),
    90, **90**
knifefishes, 22, 24; electric, 37; gymnotid, 62, 99
Koi. *See* carps: Koi

labyrinth fishes, 103, 146
lampreys, 2, 9, 12, 37, 80, 82, 94
lancelets, 8, 80
lanternfishes (*Myctophum*, Myctophidae), 13,
    24–25, 62, 65, 100
Largemouth Bass (*Micropterus salmoides*,
    Centrarchidae), 42, 43–44, 81, 90, 91, 95,
    111, 112
larvae, 34, 41, 49, 77, 78, 82, 119; as food, 87,
    91, 96; as term, 80
lateral lines, 46, 97, 99
Lepidosirenidae, 18; *Lepidosiren paradoxa*, 19
light, 24–25, 29, 30, 33, 56
lionfishes (*Pterois*, Scorpaenidae), 3, 44, 57,
    104, 111, 116, 139, 146
literature, 134–35, 139–43
litter, 124–25
livebearers. *See* reproduction/breeding
liver flukes, 114
lizardfishes, 54, 89
loaches (Cobitoidea), 8, 59, 67, 74, 82, 103, 146
Loch Ness, Scotland, 129
loosejaws, 33
lungfishes (Dipnoi), 7, 12, 15, 18–19, 66, 80,

94, 98; African (Protopteridae), 18–19, 67;
South American, 19

mackerels and tunas (Scombridae), 2, 13, **14,**
56, 71, 144, 146
mackerels, types of: Atlantic (*Scomber scombrus*),
7; King, 4; *Scomber*, 16. *See also* tunas,
types of
Mahi-mahi/Dorado/Dolphinfish (Coryphaenidae), 2, 22, 90, 146; *Coryphaena hippurus*, 143
mahseers, 16, 130
males, 38, 39, 53, 74, 75, 80–81, 82–84; and
coloration, 28, 30, 31, 34, 35
marine fishes, 3, 5, 19–20, **21,** 74, 77, 115, 118;
and aquarium keeping, 103–4; and coloration, 34; and sex reversal, 74
marlins (Istiophoridae), 2, 40, 84, 88, 91, 95;
Black (*Istiompax indica*), 10; Blue (*Makaira
nigricans*), 10, 47, 137
menhadens (*Brevoortia*, Clupeidae), 4, **5,** 47,
47, 142
mermaids, 128-29
midshipmen. *See* toadfishes and midshipmen
Milkfish (*Chanos chanos*, Chanidae), 93, 145
minnows (Cyprinidae), **4,** 8, 13, 19, 21, 70,
135, 146; and alarm substances, 59; and
aquarium keeping, 103; and coloration, 33,
34, 35; and eggs, 78; as endangered, 118;
feeding by, 93; and food chains, 71; growth
of, 84; and hearing, 97; and migration,
61, 62; mirror-sidedness of, 26; as pets,
2; and predators, 56; and relatedness of
fish in nest, 81; and reproduction, 74; and
schooling, 43–44; and sex determination,
82; spawning by, 80; and talking, 53; teeth
of, 14, 15; and territoriality, 37; and winter
habitat, 68
minnows, types of: Cutlip (*Exoglossum maxillingua*), 89; European (*Phoxinus phoxinus*,
Cyprinidae), 45; *Garra rufa*, 3; *Labeo*, 88;
Nocomis, Semotilus, Exoglossum, 72; *Paedocypris progenetica*, 11; pike (Cyprinidae), 89
mirror-sidedness, 26–27, **28**
moray eels (Muraenidae), 20, 42, 47, 74, 91,
93, 108, 115; Green (*Gymnothorax funebris*,
Muraenidae), 138. *See also* eels
mudminnows (Umbridae), 8, 68
mudskippers (*Periopthalmus*, Gobiidae), 96
mullets, 44, 56, 80, 93
muskellunge (*Esox*, Esocidae), 112, 145
mussels, 78–79
Myleus, 14
Myllokunmingia fengjiaoa, as earliest known
fish, 9
mythology, 128–31, 136

needlefishes (Belonidae), 52, 56, 90
North American Paddlefish (*Polyodon spathula*),
94, **100**

Nurseryfish (*Kurtus gulliveri*, Kurtidae),
83, **84**

Oarfish (*Regalecus glesne*, Regalecidae), 10, 128
Orange Roughy (*Hoplostethus atlanticus*, Trachichthyidae), 30, 126, **126**
Oscar (*Astronotus*), 104, 146
osmoregulation, 19
ostariophysan fishes, 59
Ostariophysi, 8, 59
Osteoglossomorpha, 8
ostracoderms, 8–9
otophysan fishes, 97
oxygen, 17, 18, 66–67, 83, 118–19, 123

paddlefishes (Polyodontidae), 8, 12, 15, 62, 94;
Polyodon spathula, 98–99
parasites, 40, 41, 42, 69, 70, 73, 88, 114
parrotfishes (Labridae), 14, 15, 20, 44, 61, 62,
71, 72; feeding by, 93–94; and sex reversal,
74; Striped (*Scarus croicensis*), **47;** and territoriality, 39
pearlfishes (*Encheliophis*, Carapidae), 40, **41**
pencil catfishes. *See* Candiru
perches (*Perca*, Percidae), 7, 13, 16, 68, 81;
climbing, 18. *See also* Yellow Perch
pest fishes, 110–11
Petromyzontidae, 12. *See also* lampreys
pets, 2, 5, 103–5, 121, 126, 144, 146. *See also*
aquariums
Pfiesteria (feestaireya), 70
pickerels (Esocidae), 2, 8, 78, **90,** 91
pikeblennies, 38–39
pike-characins (Ctenoluciidae), 89–90, **90**
pikes, 2, 89–90, 110, 119, 145; Australian Long-
finned (*Dinolestes lewini*), 90; Northern, 14,
68, 111, 112, 135
pipefishes (Syngnathidae), 16, 55, 80, 83, **83**
piranhas (Characidae): *Brycon, Colossoma*, 71;
Serrasalmus, Pygocentrus, 14, 35, 48, 58, 88,
91, 95, 104, 112, 113
plankton, 4, 20, 88, 89, 94
plunderfishes, 69; barbeled, 100
Pollock (*Pollachius*), 46, 121
pollution, 67, 68, 69, 70, 84, 114–15, 118, 120;
and climate change, 119; and fish farming,
122; and oil spills, 120, 123
pompano, 44, 96, 108
ponyfishes (Leiognathidae), 24, 26
Ponyo (animated feature), 134
predators, 23, 33, 39, 58, 71, 84, 89, 112; avoidance of, 53–60; and coloration, 28, 29,
30, 31–32, 35, 36; eyesight of, 96; human
feeding of, 108; and intelligence, 49–50;
and migration, 61; and parenting, 83; and
pest fishes, 110; piscivores as, 89–93, 94; and
prey, 23, 29, 31–32, 33, 36, 71, 91; schooling
against, 42–44; and spawning aggregations,
45; and talking, 52, 53. *See also* fighting; food

pufferfishes: (*Tetraodon*, Tetraodontidae), 32, 58, 69, 72, 88, 112; Porcupine (*Diodon*, Diodontidae), **19,** 32, 53, 58
pupfishes (Cyprinodontidae), 5, 34, 37, **67,** 81, 93, 118

rabbitfishes, 54, 71, 78, 88, 112
ray-finned fishes (Actinopterygii), 6, 7, 8, 15
rays, 2, 12, 13, 99
razorfishes, 54
reef fishes, 29, 30, 61, 62, 88, 115, 116, 143; and predators, 50, 54, 55, 57; and spawning aggregations, 44–45, 46
reefs, 36, 42, 71. *See also* coral reefs
refuge holes, 20, 30, 38, 39, 50, 54, 56, 58, 60
religious symbol, fish as, 131-32, **133**
remoras (Echeneidae), 40, 88
reproduction/breeding, 11, 20, 37, 68, 74–86; and climate change, 119; and clutch size, 77; and coloration, 28, 30, 31, 34; and color vision, 21; and desert fishes, 66; feeding in utero, 75; and fighting, 46; and gestation periods, 78; and growth, 85; and growth lines, 85; and larvae, 80; and livebearers, 34, 59, 75–76, 78, 81, 82, 103; and migration, 62, 63; and parenting, 83–84; and relatedness of fish in nest, 80–82; and schooling, 44–45; and spawning aggregations, 44–45; and spawning checks, 85; and spawning season, 78; and spawning stupor, 45; and spawning times, 79–80; and sperm, 75, 76, 79, 81; and talking, 52, 53; and territoriality, 38; and yolk-sac larvae, 80
rivulines, 66, 78
rockfishes (Scorpaenidae), 11, 37, 75, 82, 144; Canary (*Sebastes pinniger*), 12; Pacific coast, 11; Rougheye (*S. aleutianus*), 12

Sailfish, 44, 91
salmon (*Oncorhynchus, Salmo*, Salmonidae*)*, 1–2, 20, 71, 76, 77, 78, 80, 121; and coloration, 33–34; and culture, 135; and electrosensory organs, 98; as endangered, 118; knowledge of, 144; migration, 63, 72; in myth, **137;** and native peoples, 135–36; Pacific, 2; Pink (*Onchorhynchus gorbuscha*), 69–70, **70;** and relatedness of fish in nest, 81; replacement dentition of, 95; and reproduction, 74; and sex determination, 82; Sockeye, 2; and territoriality, 37. *See also* Atlantic Salmon; Chinook Salmon; trouts and salmons
sand lances, 33, 43, 54, 71
Sarcopterygii (lobe-finned fishes), 7, 18
sardines (Clupeidae), 33, 121, 144
sargassumfish, 29, 55
Sauger. *See* Walleye and Sauger
Sawfishes (Pristidae), 129–30, **131**
scales, 13, 15–17, 27, 67, 68, 85, 86, 146; eaters of, 87–88; and mirror-sidedness, 26

scorpionfishes (*Scorpaena*, Scorpaenidae), 29, 30, 32, 55, 89, 100, 112
sculpins, 37, 55, 59, 78
seabasses (Serranidae), 2, 53, 65, 90, 95, 145–46, **147;** and migration, 61, 62; and sex determination, 82; and sex reversal, 74; and spawning, 78, 80
sea catfishes (Ariidae), 77
sea chubs (Kyphosidae), 88, 93
seadragons, 29, 55
seahorses (Syngnathidae), 16, 29, 80, 83, **83**
sea robins (Triglidae), 52, 53, 98
seatrout (*Cynoscion*; Sciaenidae), 124
sea urchins, 49, 55, 71, 72, 101, 102
shads, 93, 142
sharks, 2, 12, 58, 71, 78, 122, 126, 133; biting by, 47; and eggs, 75; and electrosensory organs, 98, 99; eyesight of, 96; feeding, 91, 94, 108; and habitat, 65; and hearing, 98; and migration, 62; placenta of, 75–76; and remoras, 40; replacement dentition of, 95; scavenging by, 101; and sharksuckers, 72; spawning by, 80; teeth of, 14; and territoriality, 37. *See also* sharks, types of
sharks, types of: angel, 54; Basking, 10, 78, 128; Blacktip, 44; Blue, 62–63; Bluntnose Sixgill Shark, 81; carpet, 29; cookie-cutter, 62, 88, 92–93; Mako, 138; Portuguese, 65; scalloped hammerhead, 61–62; Whale, 10, 63; White, 63, 75
Sheepshead (*Archosargus probatocephalus*, Sparidae), 14, 101
shiners, 35, 81
shrimpfishes (Centriscidae), 16, 55; *Centriscus scutatus,* **147**
Siamese Fighting Fish (*Betta splendens,* Anabantidae), 3, 18, 78, 146
sickness/disease, 69–71, 107, 110, 114. *See also* parasites
Siluriformes. *See* catfishes
silversides (*Menidia menidia,* Atherinopsidae), 26, 33, 43, 56, 59, 61, 71, 82
siphonophores, 24
skates (Rajidae), 2, 22, 23, 24, 82
skin, 27, 57, 58, 59, 122
slime, production of, 58
Smallmouth Bass (*Micropterus dolomieu,* Centrarchidae), 32, 90, 111, 120
smelts (Osmeridae), 43, 65, 68, 71, 80, 119
Snail Darter (*Percina tanasi,* Percidae), 101
snailfishes (Liparidae), 65; *Careproctus,* 78; *Liparis,* 17
snakeheads (Channidae), 18, 66–67, 111
Snakehead Terror (film), 111
snappers (Lutjanidae), 20, 44, 62, 78, 108, 115, 144
snooks (Centropomidae), 74, 90, 95
snorkeling, **107–8**
soapfishes (Serranidae), 58; Dr. Seuss (*Belonoperca pylei*), **147**

soles (Soleidae), 144
Southern Bluefin Tuna (*Thunnus maccoyii*), 145
spearfishes (Istiophoridae), 2, 91
spines, 55, 57–58, 112
Spiny Dogfish (*Squalus acanthias*, Squalidae), 12, 78
spiny-rayed fishes (Acanthopterygii), 8, 13, 22
Spotfin Chub (*Erimonax monachus*, Cyprinidae), **121**
squirrelfishes (Holocentridae), 20, 30, 43, 53, 54, 80, 96, 97; and migration, 61, 62
stargazers (Uranoscopidae), 54, 100; *Astroscopus*, 22, 23
sticklebacks (Gasterosteidae), 31, 34, 37, 78, 81, 82, 91
stingrays, 54, 72, 101, 112, **113**, 116
stings, 116
stonefishes (*Synanceia*, Synanceiidae), 89, 91, 112, 116
stoplight loosejaw (*Malacosteus*, Stomiidae), 33
St. Peter's Fish (*Sarotherodon galilaeus*, Cichlidae), 114
St. Pierre fish, 136
Striped Bass (*Morone saxatilis*, Moronidae), 4, 76, 146
sturgeons (Acipenseridae), 3, 8, 11, 12, 15, 59, 98, 123; Baltic (*Acipenser sturio*), 118; Beluga (*Huso huso*), 10, **11**; eyesight of, 96; feeding by, 94; Lake (*A. fulvescens*), 140; and migration, 62; overfishing of, 110, 126; Shortnose Baltic (*A. brevirostris*), 118; spawning by, 80; and talking, 53
suckers (Catostomidae), 8, 15, 34, 55, 59, 78, 97, 136; *Moxostoma*, 101
sunfishes, 14, 31, 34, 68, 78, 88, 101, 112, 145; Green (*Lepomis cyanellus*), 1; Ocean (*Mola mola*), 10, 16, 59, 77; as pests, 110–11; and predators, 54; Redbreast, 111; and relatedness of fish in nest, 81; and spawning season, 78; and territoriality, 37; and vocalizing, 53. *See also* black basses and sunfishes; Bluegill Sunfish
surfperches (Embiotocidae), 39, 41–42, 54, 61, 65, 75, 88, 96
surgeonfishes (Acanthuridae), 20, 44, 57, 71, 72, 80, 88, 93; and migration, 61, 62; replacement dentition of, 95; and territoriality, 37
sustainable fishing, 125–26. *See also* farming (aquaculture); fishing
sweepers (Pempheridae), 43, 96
The Sword and the Stone (film), 135
Swordfish (Xiphiidae), 2, 91, 144; *Xiphias gladius*, 16
swordtails (*Xiphophorus*, Poeciliidae), 29

tangs, surgeonfishes (Acanthuridae), 3
tapetails (Mirapinnidae), 147–48
Tarpon (*Megalops atlanticus*), 8, 16, 80, 96, 145

teeth, 14–15, **52**, 58, 87–88, 89, 91, 94–95, 100; pharyngeal, 87, 93, 101. *See also* anatomy: jaws
Teleostomi, 2, 6–8
teleosts, 94
temperatures, 5, 82, 85–86, 118, 119, 123
Tessellated Darter, 81
tetras (Characiformes), 2–3, 82, 103, 146; Cardinal, (*Paracheirodon axelrodi*, Characidae), 28; Mexican Blind Cavefish and Mexican Tetra (*Astyanax mexicanus*, Characidae), 29, 50–51, **51**; Spraying Characin or Splash Tetra (*Copella arnoldi*, Lebiasinidae), 83–84
Threespine Stickleback, 81
tigerfish, 58, 91, 112
tilapia (Cichlidae), 3, 104, 111, 121, 145; (*Sarotherodon galilaeus galilaeus*), 131
toadfishes and midshipmen (*Opsanus*, *Porichthys*; Batrachoididae), 8, 24, 29, 53, 55, 58; Oyster Toadfish, 106
topminnow (*Kryptolebias marmoratus*, Rivulidae), 66, 118
torpedo rays, 112
torrentfish (*Cryptotora thamicola*, Balitoridae), **68**
Totoaba (*Totoaba macdonaldi*, Sciaenidae), 118
triggerfishes (Balistidae), 53, 58, 59, **59**, 62, 72, 88, 102, 146
trout (Salmonidae), 13, 37, 70, 80, 95, 96, 98, 121; Brown (*Salmo trutta*), 111, **112**; Golden (*Oncorhynchus mykiss aguabonita*), 32; Lake (*Salvelinus namaycush*), 97–98; Rainbow (*Oncorhynchus mykiss*), 106, **107**, 111, 145. *See also* salmon; trouts and salmons
trout-perches (Percopsidae), 13
trouts and salmons (Salmonidae), 2, 13, 145. *See also* salmon; trout
trumpetfish (*Aulostomus*, Aulostomidae), 36, 50, 91; *A. maculatus*, 36
trunkfishes (Ostraciidae), 16, 136
tunas (Scombridae), 2, 10, 13, 78, 90, 96, 98, 121; and flyingfishes, 22; and food chains, 71; and migration, 62, 64; and predators, 88; and reproduction, 74; and schooling, 43
tunas, types of: Bluefin (*Thunnus thynnus*), 4, 10, 44, 118, 125, 145; Dogtooth (*Gymnosarda unicolor*), **13**; Southern Bluefin (*T. maccoyii*), 145; Yellowfin (*T. albacares*), 91

Unicornfishes (*Naso*, Acanthuridae), 82

Walleye and Sauger (*Sander*, Percidae), 68, 96, 110, 144
water, 5, 17, 19–20, 29, 66, 67, 68, 97
weeverfishes, 54, 89, 112
whalefishes (Cetomimidae), 65, 147–48
wolffishes (Anarhichadidae), 14, **15**, 94, 101; Wolf-eel (*Anarrhichthys ocellatus*), 15, 80

wrasses (Labridae), 3, 20, 31, 71, 88, 146; and coloration, 34, 35; and coprophagy, 88; and ecosystems, 72; egg laying by, 78; feeding by, 101; and migration, 61, 62; and mutualism, 40; and pipette mouths, 89; and predators, 54, 55; and relatedness of fish in nest, 81; and sex determination, 82; and sex reversal, 74; and spawning aggregations, 44; and spawning season, 78; and territoriality, 37, 39

wrasses, types of, 102: Bluehead (*Thalassoma bifasciatum*), 34, 36, 143; Corkwing *(Symphodus melops)*, 42; Humphead (*Cheilinus undulatus*), 101; *Labroides*, 48, 74, 88; Senorita (*Oxyjulis californica*), 42; slingjaw *(Epibulus)*, 91

Yellow Perch (*Perca flavescens*, Percidae), 7, **16**, 46, 59, 61, 68, 78, 111, 144
Yellowtail (*Seriola lalandi*, Carangidae), 78, 145

Zebrafish (*Danio rerio*, Cyprinidae), 3, **4**
zooplanktivores, 89, 96
zooplankton, 71, 87, 94, 98